AMERICA'S BLIND SPOT

AMERICA'S BLIND SPOT

Chávez, Oil, and US Security

Andrés Cala and Michael Economides

continuum

Continuum International Publishing Group
A Bloomsbury Company

50 Bedford Square
London
WC1B 3DP

80 Maiden Lane
New York
NY 10038

www.continuumbooks.com

ISBN: HB: 978-1-4411-8669-0

Library of Congress Cataloging-in-Publication Data

Cala, Andrés.
America's blind spot : Chávez, oil, and U.S. security / Andres Cala and Michael J. Economides. – 1st ed.
p. cm.
Includes bibliographical references and index.
ISBN 978-1-4411-8669-0 (hardcover : alk. paper) – ISBN 1-4411-8669-7 (hardcover : alk. paper) 1. United States–Foreign economic relations–Venezuela. 2. Venezuela–Foreign economic relations–United States. 3. United States–Foreign economic relations–Latin America. 4. Latin America–Foreign economic relations–United States. 5. National security–United States. 6. Petroleum industry and trade–United States. I. Economides, Michael J. II. Title.

HF1456.5.V4C35 2012
337.73087–dc23

2012002892

Typeset by Newgen Imaging Systems Pvt Ltd, Chennai, India
Printed and bound in the United States of America

CONTENTS

LIST OF FIGURES

LIST OF ABBREVIATIONS

AEO	Annual Energy Outlook
ALBA	Alianza Bolivariana para los Pueblos de Nuestra América
BP	British Petroleum
CAPP	Canadian Association of Petroleum Producers
CEPR	Center for Economic and Policy Research
CGES	Center of Global Energy Studies
CITGO	CITGO Petroleum Corporation
CNOOC	China National Offshore Oil Corporation
CNPC	China National Petroleum Corporation
CSIS	Center for Strategic and International Studies
DAS	Departamento Administrativo de Seguridad
DOE	Department of Energy
ECLAC	Economic Commission for Latin America and the Caribbean
EER	Emerging Energy Research
EIA	Energy Information Administration
ENI	Ente Nazionale Idrocarburi (Italy's largest oil company)
EOR	Enhanced Oil Recovery
ETA	Euskadi Ta Askatasuna (Basque terrorist group)
FARC	Revolutionary Armed Forces of Colombia
IEA	International Energy Agency
IHS	Information Handling Services (Energy and engineering consulting firm)
IISS	International Institute for Strategic Studies
MENA	Middle East and North Africa
NATO	North Atlantic Treaty Organization
NGL	Natural Gas Liquid
OAS	Organization of American States
OECD	Organization for Economic Cooperation and Development
ONGC	Oil and Natural Gas Corporation
PDVSA	Petróleos de Venezuela SA

PEMEX	Petróleos Mexicanos
PRI	Institutional Revolutionary Party
UNASUR	Union of South American Nations
USAID	US Agency for International Development
USGS	United States Geological Survey
WEO	World Energy Outlook
WTI	West Texas Intermediate
YPF	Yacimientos Petrolíferos Fiscales

FOREWORD

By Luis E. Giusti, senior advisor of the Center for Strategic and International Studies, on top of numerous other similar roles in public and private energy and policy institutions and in academia, and former chairman and CEO of PDVSA

Latin America is a very large region extending from the Rio Grande, the southern border of the United States of America, to the Patagonia in the southern tip of South America. The region has nearly 500 million people and is endowed with abundant valuable natural resources, but it only represents 7 percent of the global economy and less than 7 percent of the global trade.

A century of complex political evolution is the main reason for many decades of poor economic performance, but most recently the explanation can be found in the contemporary regional socioeconomic evolution. Brutal regional instability starting in 1994 with the Tequila Crisis, followed by serious episodes that affected several countries throughout the decade, forced analysts and multilateral institutions to rethink strategies. The Washington Consensus, a package of macroeconomic adjustments based on the premises that stable economic growth requires balanced budgets, low inflation, deregulation, open markets, and private investments, had not addressed properly the distribution of income, especially affecting the lower classes. A general awareness then evolved among policymakers that additional reforms were needed if Latin American economies were to grow more than 6 percent a year, the rate widely believed necessary to lower the number of people living in poverty in the region.

Rejection of the Washington Consensus empowered a resurgent left and consolidated the elections of Hugo Chávez in Venezuela in 1998, Lula in Brazil in 2002, Kirchner in Argentina in 2003, and Tabaré Vázquez in Uruguay in 2004, all of them representing left-of-center coalitions and promising to undo the "neoliberal excesses" of their predecessors. They also stressed the need to reassert their nations' independence from the

United States and limit the superpowers' influence. Yet, most of these new presidents did not really deliver on their more extreme campaign promises, especially on their plans to roll back the economic reforms of the 1990s. Brazil's Lula followed an orthodox economic policy, anchored in painfully high interest rates and the active promotion of foreign investments. In Argentina, the only significant departure from the economic orthodoxy of the 1990s was the adoption of widespread price controls and a disdainful attitude toward foreign investors. Chávez's antitrade posture glosses over the reality that Venezuela enjoys a de facto free trade agreement with the United States, the largest in the region.

The responses to these political demands have also been varied. A few leaders have behaved in a traditional, populist fashion, relying on massive public spending, on keeping prices artificially low through government controls, or on scapegoating of the private sector to cement their popularity. Most, however, have implemented responsible economic governance and have shown a willingness to absorb the costs of unpopular but necessary economic policies.

Still, a second generation of reforms followed. A few of the most important reforms included the improvement of the quality of the public sector and the governance, sanctity of contracts, private property and intellectual property, fiscal strengthening, enhancement of the legal and regulatory frameworks, improvement of the performance of the financial markets, and reform of the labor markets.

Since 2000 we are witnessing the positive results. The region seems to have learned the lessons and regained its footing. A glance at the countries shows that most of them have achieved fiscal balance; their central banks have accrued comfortable levels of reserves; inflation stands at one digit (with the exceptions of Argentina and Venezuela); economic performance has been healthy during the 11 years of the twenty-first century, and in general many positive institutional reforms are taking hold.

In the summer of 2007 the financial world was shocked by the explosion of the subprime mortgage market in the United States of America, which extended like fire globally. Certain countries, like Argentina, Ecuador, and Venezuela were hit hard, but in general the region showed great resilience.

There are two main explanations for this: international conditions provided a strong base for growth for most emerging economies, and Latin American countries have generally pursued better economic policies, with lower fiscal deficits or higher surpluses, low inflation, and flexible exchange rates. As a result, the balance of payments improved and helped to increase foreign reserves and to reduce foreign borrowing. In addition, the region has diversified exports taking advantage of the new opportunities in China and India, as well as the economic recovery in Japan.

The issue of oil in the region

The rush to find more oil in the United States of America led to Texas and California and in 1900 a huge oil discovery was made in Spindletop, Texas. Following the geological trends of that discovery, the oil companies began looking south and managed to get oil rights in Mexican territory from dictator Porfirio Díaz, then president of the southern neighbor. In a few years, Mexico became the number two producer in the world and the main supplier of oil to the United States of America. But not everything was rosy because the Mexican people began rejecting the presence of American companies and denounced abuses and collusion with Porfirio Díaz. The confrontation escalated and eventually most companies had to leave, while the Mexican oil industry lost its appeal for foreign companies and their investments. Attention was then turned to Venezuela, especially after the massive blowout in 1922 of Los Barrosos-2 in the eastern coast of Lake Maracaibo, which showed the country's immense oil potential. The large oil reserves of Mexico and Venezuela triggered a gear shift in the United States of America, prioritizing the stable and abundant supply of oil from those two countries over political interests. A century later their supply remains critical to the United States of America and the global oil balance.

Since 1985 the domestic oil production of the United States of America has been declining steadily and its import needs have increased accordingly. Yet in recent years Venezuelan and Mexican exports to the United States have dropped, the shortfall being covered by larger imports from Canada.

Against that backdrop, this book goes in depth into the subject of how for many decades the United States has erred in its policies towards Latin America, and how the consequences put in jeopardy the energy security provided by the oil imports from Venezuela, especially considering the future US oil demand and shrinking oil production in Mexico.

This has been the subject of much concern, expressed by congressmen and government officials in the United States. Four years ago the Government Accountability Office of the Congress published an extensive report on the subject which concluded with an urgent call to action aimed to diffuse "unacceptable high risks of external shocks of oil supply." Obviously, within those concerns, the most important one relates to current and more importantly future oil supply from Venezuela, although also from Mexico, but for different reasons.

So what is the answer? On the basis of very well-documented analyses, Cala and Economides surrender to the critical need of Venezuelan oil, especially the very large reserves that would guarantee steady supply from that country deep into the future. In that respect they suggest "finding more creative policies that will indirectly help boost production in Venezuela," although the country would inevitably have to successfully attract more private investment.

Doubts and concerns about Latin America and Venezuela remain however in the international economic community. Reminiscences of the terrible cycles of the early 1980s and the mid-1990s added to current political concerns give cold feet to many investors. This translates into a risk premium and a cautious approach. In summary, there is a widespread view that it is a volatile region, where good margins can likely be attained in the short term, but where the risk of long-term investments may be high.

But there is an important new reality that can itself provide a partial solution. After 13 years of poor results from its oil policies, and clearly motivated by the need of increased revenues, the Venezuelan government is pursuing ways to increase production by inviting large foreign companies to participate in new developments in Venezuelan territory. The success of this initiative will not be easy because of the unilateral changes of contracts and conditions that have characterized the Chávez years. However, ENI and large Chinese companies will be engaged in important new projects that will bring new oil to the markets.

The authors suggest a new regional policy is required to balance American interests with those of a more assertive region, one with Brazil as a regional leader to facilitate closer relations, much like Turkey in the Middle East, or Japan in Asia. A more active and effective role should be assumed by Organization of American States, which has turned into an ineffective weakened bureaucratic entity.

Concerning energy security, the increase in production in Brazil (especially the future potential of the offshore discoveries), as well as in Colombia, in the United States with new discoveries and development, and in Venezuela, will bring about an important change in the oil balance in America.

In this book Andrés Cala and Michael Economides have presented us a brilliant, excellently well-documented and thorough description of the secular flaws in US policies toward Latin America (which have created the blind spot that provides its title). Despite recent efforts to improve that situation, in many ways and specifically concerning a few countries, that blind spot still exists, but things are improving.

FOREWORD

By Chen Wei Dong, Senior Vice President of China National Offshore Oil Corporation (CNOOC) and General Director of CNOOC Research Institute

This book presents an extraordinary analysis of US energy and national security in dealing with "firebrand" Venezuelan President Hugo Chávez. The Venezuelan president has made it a priority to be the thorn in the side of the United States, all while his "Bolivarian" aspirations have spread to other countries in the region, such as Bolivia, Ecuador, Nicaragua, and Honduras. His alliance with Cuba and the Castro brothers is a centerpiece of Chávez politics, and his "brotherly" ties to Iran's Mahmoud Ahmadinejad certainly do not help his image.

Needless to say, many in the US government, many members of think tanks, many conservatives headed by the Cuban expatriate community, see a red line in Chávez, both literally and figuratively. To them, his removal is in the US national interest.

Take a deep breath, according to Cala and Economides.

First, beyond the bravado and even occasional froth, Chávez has limited himself to just that. Thanks to oil exports, he still maintains an overwhelming trade volume with his "imperialist" enemy, one that dwarfs the combined trade with all his friends.

Second, the rhetoric against Chávez only strengthens him in a country and a region where populist anti-Americanism is considered by many as a virtue.

In a brilliant stroke, Cala and Economides suggest that the way to improve US national and energy security, all while neutralizing Chávez, is to adopt policies that significantly increase global oil supplies this decade, and that invariably means the American continent. At $100 oil, Chávez is a 1,000-pound gorilla. The United States will not lose its status as the regional and world superpower but will have to consider and handle the phenomena exemplified by Chávez.

The United States and Canada are priority, naturally, but Latin America stands out as a "blind spot."

The authors make a convincing argument that Latin America's untapped resources offer the single-best realistic option to push oil prices down and to strengthen a weak economic recovery. A US policy shift toward the region would indirectly attract more energy investment to its rich undeveloped crude reserves.

The book is a "tour-de-force" in understanding energy geopolitics and especially a very lucid analysis of the Western Hemisphere's socio-economic and political sensitivities.

PREFACE

Latin America: The blind spot

The United States is facing daunting national security challenges ahead as it struggles to maintain its reigning global superpower status.

The Middle East and North Africa (MENA) are in havoc. Call it Arab Spring or geopolitical maneuvering, but the region that supplies the lion's share of the world's energy is a borderline warzone.

Iran's nuclear standoff with the West is dangerously escalating; Syria is on the verge of a civil war that could draw in proxies, from Russia and Iran, to the United States and Turkey; Yemen could still fall back into its own three-way civil war; the Sunni monarchy of Bahrain—home of the US Fifth Fleet—is still struggling to contain an uprising from its majority Shiite population; American troops withdrew from Iraq, heightening the threat of reverting to sectarian civil war pitting Sunnis, Shiites, and Kurds against one another; Egypt is far from stable, and Libya will remain semichaotic for some time. And let's not forget Hezbollah, Palestine, and the existential threat of war between the region's two military powers, Iran and Israel, the first apparently moving to build nuclear weapons and the second already in possession of them. And that is a headline summary of the most pressing conflicts.

Then again, war and energy have been historically bound. World stability depends on the free-flow of MENA's oil and gas, which explains the scary brinksmanship. The recent Libyan war in fact is the likely archetype of American power projection into the future, according to the London-based International Institute for Strategic Studies (IISS). "The choice in Libya was to be a 'super-enabler' of a European operation with Arab support. These inclinations signal the dawning of a period when 'regional solutions to regional problems' becomes . . . a central aspiration of U.S. strategic policy," IISS Director-General John Chipman predicted during the presentation of the think tank's 2011 "Strategic Survey." Indeed, Western pressure on Syria is increasingly looking like the Libyan buildup to war, except this time global powers are involved.

The MENA region though is a distraction from a rising China and an increasingly assertive Russia, indisputably America's two top priorities. The two are the only countries that could eventually be in a position to challenge the United States militarily and economically. Terrorism will remain a constant threat consuming massive resources and the hotspots in Afghanistan, nuclear-armed Pakistan, Yemen, and Somalia will not quiet down anytime soon. Stakes are also high between nuclear-armed rivals Pakistan and India, the other rising power. Throw in nuclear proliferations concerns from North Korea to the Middle East, and dozens of apparently less pressing concerns around the world, whether it's the drug war or genocide in Africa, and the powder quake is complete.

And yet this plate-full of trouble is not the worst that America has faced. After all, nuclear wars have been averted, not to mention winning two World Wars and one Cold War.

What makes the current challenges different though, more threatening in fact, is the juncture. America's number one priority—at least this decade—is the economy. Everything else depends on it. A weakened economy simply makes America more vulnerable.

Economic growth in the United States is still stubbornly slow, while Europe is in recession, even now into the fifth year of a global economic crisis unprecedented since the Great Depression. While the world's richest countries struggle, the developing world is leading growth, from China and the former Soviet Union, to Latin America. The world order is reshuffling, as it has many times, but arguably this is the biggest change since World War II. America's national security now and for the next few decades will be defined by its ability to weather this storm. That will entail imposing its military supremacy in some quarters, but fundamentally recovering its economic clout quickly, especially as regional powers like Russia, China, India, Iran, Brazil, and others fill the power vacuum increasingly left behind by the United States.

That doesn't mean that the United States is in decline, but rather that many other countries are in ascent, including some that are not exactly friendly. Considering the magnitude of the ongoing economic stagnation, retrenchment is only natural. America will remain the single most powerful country for some time. Western civilization led by the United States is evolving to fix its model in crisis. It will regroup, no doubt, and come back, but when it does it will meet a new multipolar world.

Restoring America's economic prowess is vital in this juncture, even while the country continues to evolve amid an increasingly intertwined world. "Three years ago, we confronted the worst financial crisis in eight decades. That crisis proved a fact that has become clearer with each passing year—our fate is interconnected; in a global economy, nations will rise, or fall, together," President Barack Obama told the United Nations General Assembly in September 2011.

To be sure, America's ability to assert itself globally will grow more daunting for three reasons: its future is more tied to global affairs than it has ever been; the United States is facing multiple short-, middle-, and long-term threats that have forced the country to stretch its resources, and a stagnant economy is weakening its ability to deal simultaneously with multiple domestic and international issues. That is just the reality. There is a lot going on now, even more expected in the future, and simply not enough to go around.

And while much reform is unquestionably needed across the board, energy stands out because it permeates every other pressing concern. America needs a new energy policy in order to defend national security and its way of life. It's not an option but an imperative. And it must be realistic.

With that in mind, we have divided this book into three parts: the first is about America's domestic politics and geopolitics, especially in regard to Latin America; the second is about energy and related policies, and the final one is an analysis of the realistic options that the United States has to improve its energy security and its economy, and by extension to defend its national security and way of life.

US national security = energy security

Whether you know it or not, energy is as vital as water, food, and oxygen. Not in the old-fashioned, I-need-to-get-warm way, but in the I-love-to-sit-back-order-pizza-and-watch-a-movie-online kind of way. Human progress will increasingly be tied to our ability to generate and harness energy, no question.

Energy is a habit, an "addiction" that—contrary to what naïve politicians and wishful thinkers preach—cannot be kicked. In fact, we can't live without it. Everything is made with energy, even the hose that waters homegrown organic tomatoes that Occupy Wall Street protesters eat.

Yet, judging from the way our society relates to energy, it would seem electricity in our homes or gasoline for our cars will keep flowing no matter what. But that's not the case. If we want to keep our evolving quality of life, whether its internet or the neighborhood's soccer league, we are just going to have to start giving energy the same importance that we give to water.

Competition for energy sources is as old as politics, starting with fire. But in the past decade it has become increasingly clear that there is not enough *cheap* energy out there for everyone and that production costs will inevitably rise.

The Western world is currently making room for upcoming giants that will compete vigorously for what Americans take for granted, that is, secure and economic energy supplies. As long as cheap energy dwindles, governments will have to vie for their share. It's a race that involves everything from climate change to deepwater oil drilling, from technology to military force, and from geopolitics to state legislatures.

It is called energy security and it is critical to the national security of any country. And it's measured in many ways. World governments and national and multinational institutions rely on numerous sources to standardize global energy information. Most countries use their own national estimates in combination with other sources. The private industry and research centers produce independent reports. The two institutional bellwethers are the International Energy Agency (IEA), advisor to the wealthiest nations grouped under the Organization for Economic Cooperation and Development (OECD), and the US Energy Information Administration (EIA), a division of the US Department of Energy (DOE).

Here is the IEA's description of energy security:[1]

> Energy security, broadly defined, means adequate, affordable and reliable supplies of energy. It matters because energy is essential to economic growth and human development. No energy system can be entirely secure in the short term, because disruptions or shortages can arise unexpectedly, whether through sabotage, political intervention, strikes, technical failures, accidents or natural disasters. In the longer term, under-investment in energy production or transportation capacity can lead to shortages and consequently unacceptably high prices. So energy security, in practice, is best seen as a problem of risk management, that is reducing to an acceptable level the risks and consequences of disruptions and adverse long-term market trends.

---

[1]WEO 2007, p. 160.

> Short-term threats to security concern unexpected disruptions, whether of a political, technical, accidental or malevolent nature. Long-term threats relate to a lack of deliverability, caused by deliberate or unintentional under-investment in capacity. Both short-term disruptions and under-investment result in higher prices, causing hardship to consumers and harming economic prospects.

Regardless, the irreversible trend is that the developing world will consume increasingly more energy, which simply means that the United States will have to pay proportionally more for its own energy supply. China, which already uses more energy than any other country, won't replace the United States as the world's top oil and gas importer for some time. But available cheap supplies will dwindle in the process. Energy consumption in non-OECD countries will increase more than 30 percent by the end of the decade, while OECD countries will need about the same.

Growing Chinese energy demand simply upended the balance. It's not alone either. China is the biggest of the developing economies, but the trend is widespread. Emerging countries' consumption trends are approaching those of industrialized nations. Computers, cars, clothes, everything made in Asia, especially in China, will increasingly be bought by a rising global middle class, not developed countries like the United States. All that stuff requires energy to produce and run.

That means that for the remainder of the decade securing energy at prices that stimulate economic growth will get harder, especially as demand increases faster than supply. Only by 2025 is energy availability expected to improve.[2] (This will be expounded upon in Chapter 6.) So while energy won't run out any time soon, the difference between supply and demand will narrow, meaning prices will increase, and that's without taking into account unpredictable major disruption like wars, droughts, or accidents. US expenditure on energy will inevitably continue increasing throughout the decade, despite improved efficiency and domestic production, as we'll further analyze in Chapter 5, to the tune of nearly $1.4 trillion by 2020.[3]

---

[2]EIA 2011 Outlook, IEA WEO 2011, and other independent forecasts, including BP, OPEC, and others.

[3]EIA 2011 Outlook energy expenditure and energy intensity forecasts.

This is why Europe and industrialized nations in Asia have planned their energy futures with 20- and 30-year visions that are already being implemented. China, India, Brazil, and Russia are already big players in world energy markets, as consumers and producers. They all have long-term energy plans designed to run their surging economies, much like the industrialized world did in the twentieth century. Like many more in Asia and Latin America, most of that energy will go to lift millions of people out of poverty, namely reflected in souring demand of power generation and transport fuel.

It follows then that one of the greatest threats to American and Western economies is the increasing competition for affordable energy supplies. And oil is the benchmark for energy prices and will continue to be for the foreseeable future. But the not so common compounded effect of stagnant growth and high oil prices is testing the global economy. Even a European Union collapse could arguably be weathered easier than a prolonged cycle of slow economic growth and bullish oil market instability toward the $200 a barrel mark. As prices continue to climb, the United States will struggle more and more to return to robust growth, and without a solid economy all other pressing national security issues will get harder to address.

And yet, energy remains off the radar screen for most Americans, at least disproportionally so considering how transcendental it is to daily life as a silent player in all business and political decision, whether it's in a town council or the White House. Sometimes it's directly about energy, like deciding whether to open up offshore drilling after the Gulf of Mexico spill. But most of the time it's not so obvious, like jobs heading overseas, the trade deficit, or taxes. Regardless, it's always a major factor in just about any decision, even our personal ones.

Energy determines our life at least as much as any other pressing issue, if not more, yet there is little public interest or understanding of the delicate balance between how much reliable energy is available and how much we consume, and all the forces set in motion to address that very issue. Perhaps it's all those figures in alien measurements that very few people bother to understand, but the fact is that energy is an issue that nobody, whether student or retiree, mechanic or doctor, can afford to ignore.

On the surface, it's not so complex. Energy like any other commodity, say corn, for example, can be produced, distributed, and consumed in dozens of shapes and forms, from a wood stove in Africa to an ocean wave-powered generator. And there are cheaper and more expensive alternatives, depending on how it's going to be used. Prices fluctuate depending on available supply to meet demand.

On the ground, it's a very complex process set in place and improved over the centuries to ensure everyone can charge an iPod or fill up a jet ski. The developed world consumes a lot more energy than it can produce by itself. It didn't use to be that way, but it will remain so for decades. There is no way around it. And that is not necessarily a bad thing. Energy, like other commodities, is exchanged in a global, very versatile market. Energy independence is an unattainable pipedream that in any case would harm American interests. Just imagine how much work and time went into laying, and goes into maintaining, the infrastructure necessary to power a television through an outlet that is somehow connected to a power generator far, far from the screen. Or everything that involves driving to the corner gas station to fill up a tank, from finding and producing petroleum, to the millions of miles of pipelines and cargo ships that exist just to move it around all over the Earth, defying pirates, across oceans and canals, or perhaps surviving guerrilla attacks. But rest assured that energy is the one thing, along with water and food, that can determine our way of life.

We might hear about energy in the rare case of power shortages, or temporary disruptions as a result of a war in an oil-producing country or a devastating tsunami that forces the shutdown of nuclear plants. Or when gasoline, heating, or power bills increase to the point of public discontent, and politicians from across the spectrum manipulate the issue to defend ideologies that have little to do with energy.

But everyone should be concerned about a future when driving too long is prohibitively expensive, or flying is a luxury, or homes just need to be a little hotter during summer, and so on. And since a vast majority of us would not like changing our lifestyles drastically, not to mention quickly, we all have something at stake when America decides how to meet its voracious energy appetite at an affordable cost to sustain our evolving lifestyle.

However, the public debate over energy supply in much of the Western world is, in fact, misleading. By the sound of it, replacing one source of energy with another is as simple as approving a bill in Congress. But it's not. It will take decades to even dent overall demand for fossil fuels, not to mention trillions of dollars to develop the infrastructure. A transition out of fossil fuels, especially oil, takes place over generations, not years.

Debates in Washington and other capitals are really about what the future will be in over two or three decades, not in the next election.

To makes things worse, politics get in the way. And energy gets caught up in a confusing mesh of issues, ideologies, conflicting arguments, and ultimately misinformation and manipulation.

Renewable irrelevance

Before we delve further, we'll address a significant, yet widely popular, misconception simply because there is a temptation in American and Western societies to misjudge energy as choice.

For over a decade policymakers have led us to believe that a green revolution powered by renewable energy is required to improve energy security, to cut energy spending, and to avert global warming. When the global economic crisis started in 2008, rich countries decided to direct massive financial and political capital to catalyze this revolution, promising to create jobs, to cut oil import costs, and to save the world from an Armageddon.

But the opposite happened. Countries leading the renewable industry are backtracking. Let's assume that the pre-crisis world actually had the economic muscle to absorb the costs of an energy transition, but to insist on it now is simply irresponsible, or at least naïve.

The European Union, still the undisputed leader in renewable power, has a legally binding target of generating 20 percent of its primary energy through renewable sources. But aside from the investment required to increase generation capacity by installing wind and solar power, EU countries require 1 trillion Euros—$1.4 trillion—investment[4] to build a grid and accompanying infrastructure.

That is utterly inconsistent with Europe's reality. Greece has already technically defaulted on its debt, at least according to rating agencies. Ireland and Portugal have a high risk of being next, and to top it all off, Spain and Italy could also be dragged into insolvency. European economic growth will stagnate in 2012 and 2013, at best. That goes without considering the worst-case scenario, which is a breakup of the

---

[4]EU Commission, 2010: http://ec.europa.eu/energy/publications/doc/2011_Energy2020_En.pdf

European Union. Europe has no money to spare, plain and simple. Not Germany, not France, not the United Kingdom.

So how can Europe conceivably expect to meet its renewable target, especially considering the staggering investment it requires is supposed to be borrowed?

"Foremost among the challenges facing renewable energy is that somewhat bleak fiscal forecast for the US and Europe. The current environment of budget austerity and sovereign debt crises forces fiscal choices that make it difficult to sustain subsidies for renewable energy (which are still required despite improving project economics)," according to the Standard & Poor's (S&P) report published in September 2011.

The alternative energy research division of Information Handling Services (IHS), an independent consultancy and a global bellwether in the energy industry, agrees. "The EU's commitment to 2020 renewables targets remains strong, but the majority of countries will miss their targets due to cuts in government renewable budgets and transmission and permitting constraints," IHS Emerging Energy Research said in a July 2011 report.

The three renewable energy celebrities Germany, Spain, and Italy have already cut generous subsidies. The US government too has backtracked on its own commitments. "The experience in Spain, the Czech Republic, and Italy indicate the risks of a backlash to generous subsidies," S&P said.

Renewable energy annual growth is now forecast to slow to 9 percent globally at least until 2015, down from 21 percent in 2009 and double-digit growth throughout most of past decade. Solar capacity additions likely peaked in 2010, according to IHS. That is despite costs falling drastically, as much as by 50 percent from earlier highs in the past decade.

Global renewable investment, excluding large hydroelectric power, increased more than 40 percent in 2010 to around $130 billion globally from 2008, according to IHS. But it will fall for the first time in 2011 and not recover its peak until 2015.

Wind power capacity additions, which contribute around two-thirds of all renewables, last year fell for the first time as declines in the United States could not be offset by growth in Europe and China, according to IHS. In fact, overall additions will not "surpass the level of renewable capacity additions reached in 2009 before 2020," and annual growth of 21 percent in 2009 will drop to 9 percent annually for the next 15 years.

Michael Lind is Policy Director of the Economic Growth Program at the New America Foundation. In May 2011 he explained in an article published in *Salon* the irrelevance of renewable power:

> The arguments for converting the economy to wind, solar and biomass energy have collapsed. The date of depletion of fossil fuels has been pushed back into the future by centuries—or millennia.
>
> The scenarios with the most catastrophic outcomes of global warming are low probability outcomes—a fact that explains why the world's governments in practice treat reducing CO_2 emissions as a low priority, despite paying lip service to it. But even if the worst outcomes were likely, the rational response would not be a conversion to wind and solar power but a massive build-out of nuclear power. Nuclear energy already provides around 13–14 percent of the world's electricity and nearly 3 percent of global final energy consumption, while wind, solar and geothermal power combined account for less than one percent of global final energy consumption.
>
> The disasters at Chernobyl and Fukushima have dramatized the real but limited and localized dangers of nuclear energy. While their initial costs are high, nuclear power plants generate vast amounts of cheap electricity—and no greenhouse gases. If runaway global warming were a clear and present danger rather than a low probability, then the problems of nuclear waste disposal and occasional local disasters would be minor compared to the benefits to the climate of switching from coal to nuclear power.
>
> The arguments for converting the U.S. economy to wind, solar and biomass energy have collapsed. The abundance and geographic diversity of fossil fuels made possible by technology in time will reduce the dependence of the U.S. on particular foreign energy exporters, eliminating the national security argument for renewable energy. Without massive, permanent government subsidies or equally massive penalty taxes imposed on inexpensive fossil fuels like shale gas, wind power and solar power may never be able to compete. For that reason, some Greens hope to shut down shale gas and gas hydrate production in advance. In their haste, however, many Greens have hyped studies that turned out to be erroneous.
>
> Eventually civilization may well run out of natural gas and other fossil fuels that are recoverable at a reasonable cost, and may be forced to switch permanently to other sources of energy. These are more

likely to be nuclear fission or nuclear fusion than solar or wind power, which will be as weak, diffuse and intermittent a thousand years from now as they are today. But that is a problem for the inhabitants of the world of 2500 or 3000 A.D.

In the meantime, it appears that the prophets of an age of renewable energy following Peak Oil got things backwards. We may be living in the era of Peak Renewables, which will be followed by a very long Age of Fossil Fuels that has only just begun.

To be sure, no matter how much renewable or nuclear energy capacity is developed, coal, gas, and petroleum will simply supply the lion's share of the world's energy at least through the first half of the twenty-first century, as we'll show in Chapter 4. Alternatives are just that. They can't replace. They can merely add, at least for decades to come, to our energy mix.

But precisely because it takes so long to implement life-altering consumption trends, America must decide as soon as possible what it will do, mainly because other countries are certainly not waiting.

Oil is king

Oil inevitably is dragged into politics because it is so critical to any country's security and because it attracts more money than just about any other market, all things considered. A nation's ability to exert power nationally, regionally, or globally is directly related to its energy supply and the costs incurred to secure it.

But the United States, still the world's undisputed super power, economically, diplomatically, and militarily, has a blind spot in its energy policy and by extension on its national security, as we'll analyze in Chapter 5. Not enough is being done to preempt the increasing competition for energy resources in the next few decades, or the risk that high and volatile energy prices could pose to world stability and economic growth.

For the United States, the world's biggest oil consumer and the third-largest producer, the top priority in terms of energy security will continue to be protecting the existing market-based flow of oil in which the balance of supply and demand are the ultimate regulators. That requires mostly a militarily deterrent. But a myriad of economic, technical, and geopolitical factors will limit global oil production growth this decade.

The American dream was fueled by oil and sustaining that dream will become increasingly expensive, unless global oil output rises significantly. How American leaders manage this reality now will determine the country's global standing for years to come.

The single most important driver of shifting dynamics in world oil markets is China, an issue we'll dissect in Chapters 6 and 7. It has a voracious energy appetite, one still small compared to major consumers like the United States, but growing so fast that it alone will continue to account for most of the world demand growth through this decade and more than likely the next.

We'll also analyze the oil market's inability to fully regulate itself. There is plenty of oil to produce globally, no doubt, but a combination of economic and geopolitical factors appear set to continue limiting the world's ability to keep up with rising demand in emerging countries.

In Chapter 6 we'll show why the reign of oil as the undisputed king, which started in the middle of the last century, is far from over. The era of oil consumers was replaced in this century with the era of oil producers. Slightly more than half of the world's production through 2020 will come from countries outside the Organization of Petroleum Exporting Countries (OPEC), but the overall share controlled by the cartel will increase gradually. Non-OPEC capacity will continue pumping at full speed, driven by high prices, namely in Canada, the United States, and developing non-OPEC countries like Brazil, as well as Caspian and African countries.

The options are limited and resource nationalism is increasingly the weapon of choice, especially for those countries trying to check American might. The United States needs more oil, especially outside the MENA region, in order for prices to fall. It's an economic and geopolitical imperative. How much? A sustainable range for both consumers and producers is around \$100–\$120 a barrel. The lower range is realistic because it's the benchmark set by Saudi Arabia, the ultimate decider. But geopolitical instability—from Iran to Sudan—is increasingly threatening the world's ability to produce oil at sustainable prices. And even powerful Saudi Arabia will eventually be tested unless the world's potential is released.

IHS expects prices to rise to \$140 a barrel by 2015 and decrease gradually by 2025.[5] The United States and OECD countries will be hard

[5]IHS EER Global Renewable Energy Forecast 2011–2025.

pressed to cope economically with the higher range forecasts amid this growth slump. Stagflation could plague developed economies with all the implied consequences. In Chapters 4, 5, and 6 we'll explore oil prices.

Non-OPEC production growth is already constrained and working at its maximum. OPEC producers, on the other hand, are either unable or unwilling to boost production, as illustrated during the Libyan war. In the six months that Libyan production remained shut-in, only Saudi Arabia and Angola were able to significantly increase production to offset the decline, in the Saudi case at a 30-year production record. And even as Libyan output recovers to prewar highs, it's clear that OPEC's ability to keep up with increasing global demand, despite flat output in OECD countries, is constrained.

Oil prices compounded fears of a double-dip recession during 2011 and threaten to continue doing so throughout 2012 and beyond, depending on OECD countries' ability to deal with the looming debt crisis in Europe and the United States. If Libyan production was destabilizing, just imagine what would happen if Iran's ability to produce was threatened by domestic instability or a broader military confrontation in the Middle East triggered in Syria. Or if Iraq reverts into sectarian war and all the oil that is expected from its fields cannot be pumped out. Or even worse, if a regional war breaks out in one of many trouble-spots.

And imagine voter discontent in Western countries against incumbents—like President Obama—as they face rising oil prices at this economic juncture. Energy security should be a priority simply for cynical political goals.

Furthermore, increasing global oil production is foremost a domestic economic affair. The United States has entered a period of slow growth after narrowly averting a full blown economic depression. It is weighed down by crippling debt and an unbalanced foreign trade partially driven by oil import dependence at high oil prices. America's ability to hold its ground in world affairs depends on how its economy is restructured in the aftermath of the crisis. And, regardless of the scenario, that entails improving energy security.

Why Latin America

Global oil production is already expected to increase at unprecedented speed. By 2020, the total global liquids supply, which includes oil, its derivatives, as well as biofuels, will increase by around 10 percent.[6] To put that into context, that is equal to about half of America's current demand, just a bit less than Saudi Arabia's total output, or about what the United States and Mexico produce combined. It's a lot.

But just about all of that will be consumed by non-OECD countries. The Untied States in fact is likely to see a decrease in net oil imports by the end of the decade, which currently account for about 20 percent of the country's total energy use. Where the oil and other liquids come from is a component of America's energy security, but it's not the most important one. After all, oil can be bought openly in the market. If refiners need oil from somewhere else, they'll just bid up the price and buy it, like Chinese refiners do when they need to. There are refinery specifications to take into account making the buying process a bit more complicated, but easily manageable. No oil is irreplaceable.

In any case, Obama and bipartisan political leaders agree that Canada will remain America's first choice when it comes to energy supplies. The more America's northern neighbor is able to boost its exports, the more the United States will improve its energy and national security. That said, politics can trump common sense. For example, the Obama administration's decision to deny approval of a pipeline expansion from Canada in effect could slowdown its neighbor's oil development plans, which is in nobody's interest. It was a political, not a technical decision ahead of 2012 US presidential elections. And even without this new delay, Canada's ability to produce a lot more beyond the 40 percent it is already planning is constrained. Only Iraq, with its very particular circumstances, could outdo that. In other words, Canada is already factored in as a secure source and any surprise would be on the downside.

The only region that could significantly increase oil output on top of market expectations is Latin America, which has the world's second biggest oil reserves, a fact we'll develop upon in Chapter 7. And within that, Venezuela is the only game changer. It's "the only Persian Gulf country in the continent," as many commonly describe it. Venezuela's 297 billion barrels are already the world's biggest reserves, and still growing. There is no easy answer, but a new policy approach toward Latin America is required.

And it will inevitably involve dealing with Venezuela's President Hugo Chávez, the firebrand, populist, anti-American tongue-lashing

---

[6]The IEA forecast is a little smaller than the US EIA's, but the trends are similar.

comandante that has based his entire political career on undermining US interests globally. He is also facing reelection on October 7, 2012 and he's the best placed to win, unless the cancer he was diagnosed with thwarts his plans. After four rounds of chemotherapies and almost a year of treatment in Cuba, Chávez admitted in March 2012 he had undergone yet another operation to remove a new malignant tumor. He is undergoing radiation therapy, but assured his followers his health would not affect his ability to govern. His health is an utter mystery and speculations abound, but it's clear that it's so far the biggest threat to his hold on power. He remains the frontrunner to win the next elections for another six-year term. Efforts to alienate the anti-American president have backfired and now even Colombia, the most important American regional ally, is moving to seek its own stable relations.

Is Venezuela a silver bullet for oil markets and indirectly for US energy security? Quick answer is that it could be, perhaps more so than any other country in the world, and certainly in the continent, as we'll explain in Chapter 7. A more inviting energy sector, one that attracts oil majors, could almost double its production within the decade. That would be almost twice what Libya produced before the war and more than what the United States expects Iraq to bring online this decade. Oil prices would be sure to fall.

If Latin America, led by Venezuela, gets the money and expertise necessary to achieve this goal, incremental global oil supplies by 2020 would increase by at least 30 percent, which alone would push oil prices significantly lower. Few believe Venezuela under Chávez could deliver because his policies are simply unable to attract all that is required to meet the target. The obstacle is political though. The technological challenges are surmountable. It just requires attractive investment policies from Venezuela.

One of the main obstacles though is that the United States remains bogged down in the region by expired ideological turf wars, increasingly distancing itself from Latin America, and indirectly delaying oil plans in the region by implementing policies that ultimately increase the risk premium for private investors. That is not to say that the United States is responsible. This is not a blame game. But preoccupied with domestic politics, the United States is missing out on the chance to forward both its strategic and economic interests in the region, while every other world power happily nests in what Washington once described as its "backyard." And it doesn't need to be this way.

An American strategic policy shift toward Latin America, as we intend to spell out in Chapters 8 and 9, would attract more investment and spur oil production, while strengthening its strategic standing and weakening Chávez. And best of all, it wouldn't involve sacrificing American values like defense of human rights, of democratic values, of freedom of the press.

So why not address America's blind spot in Latin America?

PART ONE

AMERICA'S FAILED RELATIONS SOUTH OF THE BORDER

1 NATIONAL SECURITY: THE IDEALIST CURSE

America's blind spot is Latin America. It's not a secret or a newsbreak. When the Soviet Union collapsed in 1989, Latin America fell off the radar screen. And to be fair, during the 1990s there was little reason to pay attention to it. It was immersed in an uncertain transition, like most of the world, and it was unclear where it was heading. For that matter, most of Asia and Eastern Europe were in a similar situation.

When Americans think of their neighbors south of the border, most imagine a confusing mesh of Spanish-speaking Banana Republics in semifeudal drug-trafficking states run by corrupt officials, despotic generals, and rich hacienda owners enslaving poor and illiterate Mayas and Incas.

But the reality is different. The 21 countries stretching from Mexico to Argentina, including most of the Caribbean, have very little in common, other than some history. Mexico, for example, doesn't share much with Brazil or Argentina. Canada and the United States are a lot more closely related. Ask a Mexican where Bolivia or Paraguay are and most will go blank. Ask them what the capital of the United States, France, or Spain is, and they won't flinch. The indifference is mutual. South Americans in general know or care very little about Central America or Mexico. Historically, the region has been anything but united. Until recently, each country sought its own destiny through bilateral relations with the West, not with its neighbors. It can be more expensive to fly directly between Colombia and Venezuela than triangulating through Miami, which is three times farther. Brazil, the biggest and most populated country, does not even speak Spanish, but Portuguese.

Perception is powerful when it comes to policy-making though. Naturally, most US policymakers are fully aware of the reality in Latin American countries, but US citizens are not. This is critical because policy toward the region is driven by domestic issues, not foreign interests. So while the White House and Congress are sufficiently proficient on Latin American issues, reality on the ground is mostly irrelevant. Washington simply is not paying attention.

The United States missed the memo on Latin America's evolution into a vital region in global affairs. Policies are still stuck addressing, at best, Hollywood-exacerbated stereotypes, but in reality are mostly driven by outdated and counterproductive Cold War misconception pressed by powerful interest groups.

It's understandable that the United States didn't consider Latin America a priority last century. But policymakers—perhaps preoccupied with the Middle East and the worst economic crisis in decades—are increasingly undermining America's security by ignoring significant strategic shifts in Latin America.

This is about being able to protect global American interests, about national security, about global terrorism, and above all about the country's economic growth that is vital to address Middle East instability, to contain a resurging Russia, and to preempt a soaring China that is not hesitating to fill the vacuum left by the United States.

To be sure, this is about America's future as a world power. And it all starts at home.

The idealist curse

The debate over how much the United States should intervene in global affairs dates back to the founding fathers, through the continent's march to independence, the age of imperialism, the World Wars I and II, and the Cold War. There are two core sides, the realist and idealist visions of foreign policy. The divide transcends party lines and the middle ground is in constant flux.

In December 2009, when Obama received the Nobel Peace Prize in Oslo, he laid out the debate: "Within America, there has long been a tension between those who describe themselves as realists or idealists—a tension that suggests a stark choice between the narrow pursuit of interests or an endless campaign to impose our values around the world."

Realists believe American intervention in foreign affairs should be driven by a pragmatic defense of American interests. Idealists believe that what's best for America is best for the world and that the country's foreign policy should literally be designed to defend democracy, human rights, and basic freedoms on a global scale. Realists prioritize the defense of American interests, even if at times that means disregarding some idealist values. Idealists condition foreign policy to moral imperatives, even if America's interests are not best served. Realists can deal with all sorts of shades of gray as long as it's in America's interest and that basic lines—like genocide—are not crossed. The moral bar is significantly higher for idealists who, for example, want to sanction China over its Tibet policies. Idealists are not homogenous either and different groups often defend contradictory things.

It's not an either/or choice between heartless calculations or sobbing duties. Nor is it any different from the decisions made by every person every day. As corny as it may sound, America's foreign policy is boiled down to the age-old question over how much "we should think with our hearts."

In practice though, foreign policy is a combination of both realism and idealism. Each circumstance is different and there is no formula, although public opinion is a decisive factor. That explains why the United States intervened in Kosovo, but not in the dozens of devastating African civil wars, and why some dictatorships are tolerated and others aren't. It's not hypocrisy, as many simplistically conclude. It's common sense. How much are American interests at stake? Is it strategically important? Do we have enough power to project? How much will it cost? Will it accomplish anything? Who are the victims and what is the target? The answers often fall prey to demagoguery and interests groups that compete to influence policies. Very rarely is the decision clear-cut, like in Kosovo. Serbs were once again ethnically cleansing its territory even after the wars in the former Yugoslavia killed at least 130,000 people. NATO's (North Atlantic Treaty Organization) intervention to stop the massacre was idealist-driven, but the fact that it was Europe and geopolitically strategic undoubtedly weighed heavily in the decision to attack. If it were all altruism, all values, this world would be a giant moral battleground.

Ronald Reagan's foreign policy, portrayed as idealist for a domestic audience, was ultimately realist, to the point that it accomplished its goal of defeating the Soviet Union. His successor George Bush "the first" took realism to its highest levels. Bill Clinton and George W. Bush "the second" were more idealists, albeit with often different visions.

Being a realist doesn't imply being indifferent to freedom, basic human rights and democratic institutions. Politics is not rational and thus expecting an inflexible standard is absurd. There is no moral contradiction. Both Republicans and Democrats defend the same basic values, but where idealists control policy-making, moral considerations weigh more heavily. But throughout history, most direct foreign impositions utterly driven by ideology have backfired. Stopping an impending massacre is a lot easier than changing a country's fate—as the Libyan case illustrates—and the question will always linger as to whether do-gooders end up causing more harm in the pursuit of virtue.

The Obama administration has been ambiguous to the point of contradiction. President Obama came to power promising a reversal of Bush's foreign policy, one based on defending American interests. In other words, a realist approach. In practice though, foreign policy has been a confusing mesh of Bill Clinton's and George W. Bush's idealism, repackaged into Obama's realism. Consider how the White House tolerates antidemocratic regimes in the Persian Gulf, like Bahrain and Saudi Arabia, but has aggressively supported pro-democracy forces in Tunisia, Egypt, Libya, and Syria.

To be fair, Obama's foreign policy had to be a continuation of Bush's era simply because he inherited the country's most pressing problems, including a financial crisis and the Iraq and Afghan wars. Whether he has managed them well is a different question, but he didn't create them. The Arab Spring was also clearly unexpected, and Europe's crisis has been a constant distraction. And he has been preoccupied domestically with economic issues, the top priority for Americans, and as a result a coherent long-term policy has been slippery, especially in dealing with China and Russia.

Still, on the most pressing issues that Obama has been personally involved in, realism has weighed more both domestically and abroad, even if Obama always tries to claim the moral upper hand. His policies at home have reinforced a liberal interpretation of moral America, but on the economy, energy, and environment he has shifted from early idealist ambitions to more realist, pragmatic solutions, sidelining both extremes of the right-left political spectrum and seeking to attract the center of both Democrats and Republicans, not always successfully.

Abroad, policies have been erratic though, exposing the internal frictions between idealists and realists in the administration. For example, Obama only sided with Egyptian protesters once it became clear that

Hosni Mubarak's government was going to fall. And he never supported regime change, but rather an organized transition. Israel, Saudi Arabia, and certainly many in the United States advised Obama to support Mubarak no matter what because the alternative would probably be worse. That indeed may have been the case, but Obama had little choice. He only supported Mubarak's removal once it appeared inevitable and only if he was replaced by a caretaker government that left the regime intact. Obama wanted to be on the right side of history to have a greater influence in the transition. It's a realist approach, but Obama has made a point of selling it as an idealist one. Some call it the Obama Doctrine, perhaps a bit prematurely.

A similar argument could be made about the Libyan war. Certainly there was little direct American interest in Libya, and indeed the Obama administration was initially against an armed intervention, even a limited no-fly zone. But France and the United Kingdom, for their own interests, were gung-ho about attacking strongman Muammar Gaddafi. On the surface, the United States was perhaps the only one that could actually justify an idealist motivation—that is, protecting civilians. Other countries like France and the United Kingdom had a more realist motivation in the form of economic interests, mostly oil. And Italy wasn't initially on board for a similar reason—too much oil. But once a French and British attack on Gaddafi seemed inevitable, Obama joined in, despite the Pentagon's advice. The US involvement though was a pragmatic decision in response to fast-moving developments on the ground, both in Libya and Europe. Washington wanted to avoid a prolonged civil war that would have kept oil prices high while opening the door to Al Qaida to settle in Libya, not to mention lingering doubts over the military capability of European countries. But of course it was all packaged as a defense of Libyan civilians.

The same can be said about the unfolding government repression in Syria. The Obama administration dragged the country in when it sided with rebels seeking to depose the government. And while the ultimate idealist motivation is to stop the killing, Obama is seizing on perhaps the best chance America has had in decades to shift the military balance in the Middle East. By targeting Syria's leadership and promoting regime change, Iran's overarching influence in Syria, Lebanon, and Palestine could be seriously curtailed, and Tehran's most formidable weapon—its asymmetric deterrent in the form of Hezbollah and Hamas—could be blunted. So yes, removing from power a ruthless dictator that is killing thousands of his citizens is morally justified and certainly makes for good media bites. But the ultimate goal is a realist one. It's a risky move that could just as well reinforce Iran if things play out to its advantage, but Obama might be weighing the decision against a credible Israeli threat of attacking its foe. If it works, it could well prevent a larger regional war by forcing Iranian military planners to retrench. Indeed, a wolf in sheep's clothing.

Policy toward Latin America is an utter failure. Instead of upgrading relations with Latin America, the White House passed down the responsibility to Secretary of State Hillary Clinton, who has pursued an idealist, counterproductive agenda. The region is still considered a second-tier priority in Washington, a big and costly misreading. And Clinton's approach is making it worse. It's like an extension of her husband's administration, which not only expired over a decade ago, but didn't work in the first place. If during the Clinton years relations with Latin America were driven by Democrat idealists, and in the Bush years by Republican idealists, in the Obama administration, which was supposed to be realist, there has been a mix of inherited Bush policies with an idealist Democrat twist.

In the aftermath of the Honduran military coup, for example, which we'll analyze in depth in Chapter 8, the State Department prioritized containing Chávez's expansion of Bolivarianism, which in policy-making circles of the United States is equated to authoritarianism. The realist response would have been to side with the rest of Latin America in the defense of institutional legitimacy. Instead, the idealists concluded that confronting the anti-American movement was more important, even if it meant ignoring a military coup.

The pillars of American policy toward the region are still based on the idealist policies of Bill Clinton. Free trade policy, which better managed could be the vehicle for American interests in the region, continues to be conditioned by noneconomic considerations. Colombia is the best example. In the past, the country has undeniably been at least complicit in gross human rights violations that have mostly remained unpunished. As a result, the White House and Democrat Senate held up ratifying the Free Trade Agreement (FTA) signed between Bush and Colombian President Álvaro Uribe in 2006. So while Chávez was imposing an embargo on Colombian goods, the United States was refusing to open new commerce routes for its stalwart ally, while signing other FTAs with Oman and Bahrain, not exactly exemplary when it comes to respecting human rights.

Did holding up Colombia's FTA push the country closer to respecting human rights? Unlikely. Colombia's record has indeed improved remarkably, but as part of the country's transition in a postwar scenario, and in any case only under the new administration of President Juan Manual Santos who took office in the end of 2010. If anything, American military aid had a bigger role in pacifying Colombia. The Democrat firewall did however undermine America's relation with its closest ally in South America. That is, idealist foreign policy undermined Colombia's embrace of free trade that the United States trumpets. Obama had to be personally persuaded by President Santos, and only in 2011, that America's FTA policy was simply backfiring, and in the process hurting Colombia's economy, its ties to the United States, and the very reforms designed to improve human rights. The FTA was ultimately ratified in October 2011 and will be implemented in 2012.

The same case can be made about the drug war that started two decades ago under Bush "the first." Idealist-driven US policymakers for decades have simply sought to show their constituents they are doing something about the drug trade. Most Americans don't actually get involved in the results, as long as there is something to show for it.

But in Latin America, the policy has been very unpopular and a constant source of friction. In fact, leaders of Colombia, Brazil, Mexico, and of most Central American countries in 2012 proposed opening international negotiations to somehow regulate the illegal drug trade, to somehow legalize it, defying American reluctance to even consider the issue. It's not a new proposal, but it's the most direct defiance of America's war on drugs. Not surprisingly, the Obama administration said it wasn't ready to discuss any legalization.

The United States insists on an eradication policy that year after year yields little results other than an expensive bill for American taxpayers and hundreds of dead in drug exporting countries. But policymakers continue to prioritize idealist domestic political concerns. Obama struck a realist tone at the beginning, indicating his focus would be on addressing economic factors in drug producing countries that foster the drug trade, and on curbing demand at home. But managed by the State Department and the Pentagon, the military and security component that is so unpopular in Latin America was only beefed up, and the failed drug war continues to rage.

Obama also spoke about a realist external immigration policy, but like Bush "the second," the focus has been on short-term repatriations and security. The United States showers Central America and Mexico with money for border control, but it pays little attention to the underlying structural causes for migrants wanting to come to the United States or for peasants to get involved in the narcotic trade in the first place. And in the meantime, illegal immigration remains unscathed and drug lords grow ever more powerful.

Venezuela and Cuba are of course the best examples of idealist versus realist policies. Obama also spoke of a fresh start, one led by soft power, a realist policy. Cuba has endured a US embargo for half a century, but there have been little democratic gains in the island to show for it. The recent reforms that Raul Castro has set in place were triggered by domestic economic and political concerns, not the embargo. And while European, Asian, and Latin American firms have been happy to get a head start in the island's slow pro-market reforms, American companies have remained—to their loss—sidelined, from oil to tourism. In the process, the United States has lost any real ability to influence Cuba.

United States policies toward the island are also one of the main reasons the rest of Latin America is distancing itself from Washington. The rest of the continent—Canada included—wants a more relaxed approach to influence the Cuban transition, and the United States is alone in demanding isolation.

What is more baffling is that relations with Venezuela increasingly seem to be heading in the same direction. Idealist policies this decade of both Republican and Democrat administrations under Bush and Obama are driving relations to a standoff with the cheerful support of Chávez, who naturally benefits from this.

A veteran of the inner workings of US energy policy is David Pumphrey. Today he is a senior fellow and Deputy Director of The Energy Program of the Center for Strategic and International Studies (CSIS), a Washington D.C. think tank. Before this, he was Deputy Assistant Secretary for International Energy Cooperation at the DOE, where he was responsible for strategic development and implementation to strengthen US energy security, engaging key players, including China, India, Canada, Mexico, Russia, Saudi Arabia, and the European Union. He has represented the United States in the IEA's technical committees, in the energy group of the Asia Pacific Economic Cooperation forum, and in energy negotiations of the US-Canada FTA and the North American Free Trade Agreement (NAFTA). He addressed US energy policy during an interview in Washington:

> The biggest problem is that Chávez is not important enough for Americans to care what happens to him. What they care about is whether gasoline prices are high. Influence groups, small influential groups, say it's not right to sacrifice values to get more production out of there. And the bulk of Americans don't care. Chávez is now a major factor in the [anti-Castro Cuban American community]. They've adopted Chávez as a pseudo Castro.

It's not just Venezuela and Cuba. The United States has grossly misread the evolution of Latin America, which most economists agree and investors believe is set for a long-term growth cycle. The resources and middle classes are there to finally speak of the prospect of long-term pro-market economic continuity and political stability, even if not as prescribed by the United States. Brazil is already a global powerhouse; Chile can't be judged as an exception any more, as it was for some time; Colombia and Peru have gradually stabilized, albeit slowly, and are now the rising stars; Argentina is getting back on its feet; Mexico and Central America continue to be bogged down by drug wars and political stagnation, hurting economic activity and foreign investment. But Latin America is mostly on solid footing, despite political shifts throughout the region from left to right to center and back.

The United States remains the most important player businesswise, but corporate risk aversion has increased. Economic ties, whether in terms of trade or investment, are ultimately the most important foreign policy vehicle. There are a myriad of reasons to explain why American investors prefer other destinations and many are justified. But the government-fueled political risk factor, which appears to weigh on many corporate decisions, is overblown. The fact that other global players have quickly and happily moved in where American companies have lost interest is proof enough that whatever problems the region has—and it has plenty of them—Latin America is vital to world affairs in the future. Global players with no geographic ties and even less cultural affinity have concluded there is money to be made and geopolitical ties to be strengthened, so why is it that American investment is decreasing, not increasing?

John Negroponte is a veteran of US diplomatic and intelligence communities, with over half a century of experience in Washington's decision-making process and he's one of the architects of US modern history, playing critical roles in Latin America, Iraq, and American foreign policy and national security. A London-born American of Greek descent, he is now a scholar, but has served as Deputy National Security Advisor to Ronald Reagan, Deputy Secretary of State, envoy to the United Nations, the first ambassador to Iraq after Saddam Hussein's downfall, and Director of National Intelligence.

During an interview in Washington, he dismissed Latin American criticism as unwarranted.

> I think sometimes we don't get enough credit for some of the things we have done. NAFTA, the Central American Free Trade Agreement, the treaty with Peru, and now the FTAs with Colombia and Panama.
>
> Trade is very high on our agenda. Very often we get compared with what happens on the margin. If a couple of years go by without a big ticket investment, and China comes in with a few billion, people say uuhh, aahh.

That may be true, but the inroads made by China and other powers is not on the margin, as we'll analyze throughout the book. And as long as policymakers treat Latin America as a little brother, investors will continue to be swayed by an incorrect geopolitical assessment. Perhaps some concerns are well founded and instability is around the corner, but the rest of the world doesn't seem to agree. The region's resilience, politically and economically, during the economic crisis should dispel at least some of those fears. And perhaps then a more realist, pragmatic policy toward the region will emerge to the benefit of America's long-term interests and security.

The realist version of Latin America

What would be a realist take of Latin America now and into the future?

Mauricio Cárdenas, who was senior fellow and director of the Latin America Initiative at the Brookings Institution and an influential policy expert on Latin America before being appointed Colombia's energy minister in September 2011, wrote in an article ahead of Obama's trip to Brazil in the spring of 2011:

> Other powers have begun to take notice. China, for one, seems to have a strong strategic interest in a region where the U.S. is losing influence.

China is Brazil and Chile's main trading partner, and in 2009 and 2010, the China Development Bank agreed to lend more than $35 billion to borrowers in Argentina, Bolivia, Brazil, Ecuador, and Venezuela (mostly under "loans-for-oil" arrangements). This is three times what the Inter-American Development Bank approves every year for the region as a whole.

The global recession was a small bump in the road for most of Latin America. Today, the region is growing at an average of 5 percent per year, inflation is in single digits and fiscal deficits are small. Public debt as a share of GDP is much lower than in the developed world. Chile and Peru are the two countries that stand out in terms of economic performance, but considerable success is also apparent in Brazil, Colombia, Mexico, Panama, and Uruguay. Rating agencies have granted all of them investment-grade status, which means that the risk of a default is extremely low. (Argentina and Venezuela are the two salient exceptions in the region.) Center-left governments have made macroeconomic stability a pillar of their economic strategies.

In reality, most countries—with the notable exceptions of Bolivia, Cuba, and Nicaragua—have definitively rejected Chávez's "21st Century Socialism," which is based on heavy state intervention, forced nationalizations, and fiscal profligacy. The disastrous economic consequences in Venezuela are visible. Fewer and fewer countries are tempted by the populist rhetoric and the attacks on private enterprise.

We estimate that by 2020 10 percent of Latin America's population will enter the global middle class, bringing nearly 60 million individuals up to the same level of income as lower-middle-class citizens in such European countries as Portugal and Italy.

Also, the Latin American consumer market is by no means irrelevant for US companies. With only 40 percent of China's population, Latin America's average per capita income is twice that of China's.

The region buys goods and services that generate jobs in the United States. Mexico is the second-largest market for US exports, Brazil the eighth, and Colombia the twentieth—even without the passage of the pending free-trade agreement. Their combined imports from the United States in 2010 exceeded $210 billion, which represent thousands of jobs in America, especially in the manufacturing sector.

On the political front, the United States must also understand that Latin America not only rejected Chávez's Socialism, but also military coups, even if it is to remove an anti-American leader. The region's mainstream regimes and elites have broadly concluded that economic development is a long-term priority that requires political stability, as well as reform of the exclusionary political and economic models that so easily fall prey to populist movements like Chavismo. Latin America prefers coexisting in the short term with Bolivarian groups while their appeal expires. Confronting Chavismo head-on is unnecessary because the movement has climaxed already and doing so will only embolden antidemocratic forces, from paramilitaries and drug traffickers, to reactionary and authoritarian wannabes.

Washington must also understand that Brazil, not the United States, is the model the region is following. Brazil's political and economic evolution—based on former President Lula da Silva's recipe that balances economic growth and private industry with economic devolution—has been formidable. The Lula model has delivered sustainable economic growth, democracy, and stability, while lifting millions out of poverty. And several Latin American leaders have openly adhered themselves to the model, including the presidents of Peru and Colombia. This is a good thing, of course.

Economically, most of Latin America has concluded that pro-market forces are better at managing than statist ones. The opposing visions are no longer embodied in Cuba versus Chile, but more like Venezuela versus Colombia. Whatever the ideology, Latin America needs investment and foreign expertise to materialize its potential, so even the most rhetorically anti-capitalists, like Venezuela and Cuba, acknowledge foreign private investment is essential.

In 2011, the United States exported $351 billion worth of goods and services to Latin America, compared to almost $104 billion to China and $268 billion to the European Union.[1] The region also supplies more oil to the United States than any other, including the Middle East.

Latin America's empowerment has led most countries in the region to demand a new relationship with the United States, bilaterally and regionally. Washington though doesn't appear ready. It still treats the region, Brazil included, as a raggedy group of apprentices. And so Latin

[1]2011 US Census, International Trade in Goods and Services.

America decided to move on and for some time has been seeking new partnerships regionally and globally, especially with China.

The new reality was best illustrated in the April 2012 Summit of the Americas held in Colombia. Countries could not agree on a final declaration and Obama came home with little more than a scolding. Latin America demanded an overhaul of continental relations, from the drug war to Cuba, but the United States opted to defend its standalone idealist policy.

There is still ample room for focalized instability in Latin America. The economic disparity, compounded by increased awareness, could easily result in turmoil in several countries, especially in the smaller, poorer ones. But even in those cases, the more likely scenario is of a short return to stability, as the Honduran case illustrates.

Brazil, under Dulma Rousseff's presidency, has consolidated Lula's center-left model, with a slight slant to the center. Chile has long firmed up its economic and political stability. Argentina, which extended the neopopulist resurrection of Peronism under President Cristina Fernández de Kirschner, will gradually continue to find a middle ground between pro-market and statist forces, even if it takes longer than others. The same can be said of Uruguay and Paraguay. Peru and Colombia, which have managed to more or less preserve basic democratic institutions over the past decades, without carrying out any real economic reform to redistribute wealth, are already immersed in a slow transition from right-of-center to center precisely after realizing that their exclusionary model was as unsustainable in the long term as Chavismo. Central America and the Caribbean, which are significantly more dependent on Mexico and the United States, will first have to extricate the power of drug mafias before they address anything else, and thus their stability is more uncertain.

The Bolivarian countries are more unstable because they have yet to consolidate the regime changes that brought them to power. Bolivia and Nicaragua are the most precarious because they lack the oil wealth of Venezuela or even Ecuador to finance the massive public spending that their populist governments require. Still, Nicaragua in fact reelected one of Chávez's staunchest allies Daniel Ortega in November 2011 in a vote that was criticized by international observers. Ecuador, with Rafael Correa, also appears to be following its own model. In fact, Ecuador in September reestablished relation with the United States after almost six months without an ambassador.

As for Venezuela, Chávez moved up elections scheduled for December 2012 by three months to October 7. His health aside, there seems to be little stopping Chávez from winning another reelection, even if he has shed some of his popularity. A "baseball-sized" malign tumor somewhere in the pelvis though changes all bets on the table. In September 2011 Chávez said his fourth chemotherapy treatment in Cuba was his last and that he would die old. But in March 2012—seven months before elections—Chávez had a second malignant tumor removed and will require radiation therapy as he struggles against his cancer. The seriousness of his condition will remain a State secret until he decides otherwise. Chávez's health though is understandably raising doubts about his ability to govern, despite passionate assurances that he will lead his Revolution for years to come. That said, his cancer struggle is also favoring him politically as a sympathy factor. Whatever the truth, Venezuela is openly debating a future without him at the helm. Chávez doesn't appear to be grooming a successor or preparing a transition. Perhaps he's waiting until after the election to get treatment in order to secure his party's hold on power until 2019.

His health, in any case, should be irrelevant to American policymakers when it comes to oil production, as we'll analyze in Chapter 9. Regardless who wins the elections, any government will have little choice but to increase oil output, which implies attracting foreign invest—even if it's not Western. Policymakers can chose to continue pressuring Chávez on idealist grounds, but they have to stop targeting Venezuela's oil industry because it hurts the United States and benefits every other anti-American oil producer.

Furthermore, even if Chávez remains in power until 2019, Venezuela will have no choice but to reform its revolution in order to render it economically and politically sustainable. Chávez's perpetuation in power depends on correcting many of his economic policies, including his hostility toward foreign investment, even if he remains rhetorically anti-American at the core. China, which has become Chávez's economic lifeline, is demanding the same that the United States is from Venezuela's oil industry. Either Chávez fails because he refuses to change and his support base erodes, or he implements more effective models. Either way, Chávez does not have a repressive hold on power, like Gaddafi, the Ayatollahs, or the Castro brothers. His coalition is fragile as it is. Without popular support, he's doomed. And Venezuelans are already demanding reforms, even from within his powerbase.

Henrique Capriles, his opponent in upcoming elections, is not the right-wing, neoliberal, rich elite that Chávez has made a habit of bashing. He is a left-of-center, Lula wannabe who is proposing to extend many of the Chavismo social gains, but without resorting to the political extremes, unchecked authority, and anti-Americanism that Chávez spouses.

Moreover, it's all just part of the growing pains that Latin America is experiencing. Chávez is to Venezuela what Uribe is to Colombia, what Kirchner is to Argentina. Every country has had its own mix of corrupt elites, self-defeating populist alternatives, authoritarianism, and so on. Venezuela is just taking longer to find a viable middle ground.

The decade-old Latin American transition toward a center-left model and away from an unchecked neoliberal past will not be reversed, as we'll analyze in the next chapter. The region's newly found stability is the result of a long-overdue economic inclusionary wave that is lifting millions out of poverty and into a middle class that has long existed in developed economies like the United States. In other words, Latin America is simply transitioning from developing to developed, albeit with stark differences between Nicaragua and Costa Rica, for example, not to mention Bolivia and Brazil.

The realist vision of Latin America is thus one of ongoing transition, one that started after the collapse of the Soviet Union. While risks will remain and challenges will inevitably resurface, most of Latin America is moving toward consolidating around centrist, pro-market, social-democratic models, with the exception of the Chavista block which opted for a statist model with various degrees of authoritarianism. But even this group will likely stabilize increasingly by seeking a more centrist policy, at least enough to attract foreign investment. Ecuador appears already immersed in that transition, although much remains to be seen.

Regardless, idealist-driven American policies, Republican and Democrat, have been broadly rejected. Countries want a new relation, one based on equal standing, even if the United States remains a preferred partner. And it's clearly in America's interest to deliver that reform.

2 THE BOLIVARIAN CAUDILLO AND FAILED US POLICY

Latin America has matured into a powerful region over the past two decades, and to different degrees its countries have shed their politically dependant and economically unstable pasts of dictatorships, wars, and gross human rights violations. "The 2009 global economic crisis affected Latin American and Caribbean economies severely, as demand for the region's goods and services plummeted. However, thanks to improved domestic macroeconomic management and regulation, Latin America was better equipped to tackle this crisis than ever before. Domestic demand explains at least part of the Latin American resilience," wrote Mario Pezzini, the director of the OECD's Development Center in the presentation of the Latin America Economic Outlook 2011.[1]

Latin America's combined GDP in 2011 of $5.6 trillion was slightly less than Japan's, 20 percent smaller than China's, and about a third of the European Union's,[2] although per capita spending of its nearly 600 million people is closer to that of Eastern Europeans and fast approaching developed economies' consumption patterns.[3] Economic growth was 4.3 percent in 2011 and nearly 6 percent in 2010 driven by domestic demand, and is expected to slow down to 3.3 percent in 2012.[4] The region also holds an enviable reserve of minerals, land, water, and natural resources, and its population is transforming into a robust middle class with rising number of high-income homes.

---

[1]www.latameconomy.org/fileadmin/uploads/laeo/Documents/E-book_LEo2011-EN_Entier.pdf

[2]IMF World Economic Database: www.imf.org

[3]Goldman Sachs's The Expanding Middle, 2008: www.ryanallis.com/wp-content/uploads/2008/07/expandingmiddle.pdf

[4]United Nations. World Economic Situation and Prospects 2012: www.un.org/en/development/desa/policy/wesp/wesp_current/2012wesp.pdf

In other words, it's not a potential, but a reality.

None of this has escaped China, Russia, India, Iran, Europe, and other middleweight powers. Until the 1990s, it was a proxy battlefield of Americans and Soviets, a strategic trophy rather than an investment opportunity. Now, it's a must destination for global leaders trying to make inroads, form alliances, and broaden relations with the region.

But the United States has been largely absent precisely during Latin America's formidable coming of age. As a result, it is being displaced gradually as the region's hegemonic power by Brazil, which in 2011 moved ahead of the United Kingdom to become the world's sixth largest economy. Meanwhile, the gulf with most countries in the region continues to grow. The United States is not even supporting Brazil's bid to take up a permanent seat in the UN Security Council. Instead, it is throwing its weight behind India, despite the fact Brazil is more likely to share American policy. Indeed, relations between the two continental powers increasingly appear one of competitors, or at least that was the widespread conclusion in the aftermath of an official visit from Brazilian President Rousseff to the White House in April 2012.

That is a diametrically different reality from just two decades ago when the region was still considered "America's Backyard," an infamous title it held for almost two centuries and is still used commonly in American policy-making circles.

This profound shift in relations is mostly due to the region's own political and economic maturity. Indeed, the United States was able to detach itself from its traditional paternalistic role precisely on the back of Latin American countries' growing assertiveness. Yes, the United States has been absent from the region for two decades, ever since the end of the Cold War, but that is exactly the self-determinism that Latin American countries demanded last century, and indeed what the United States has long demanded.

American leaders concluded that Latin America has its troubles, but that it's doing just fine in handling its own affairs. The United States has vastly more pressing issues to address globally and internally and its leaders felt confident low-key diplomacy was sufficient when combined with a summit here and there. The region not only sidestepped the recession, but has been a significant contributor to global growth. And not just Brazil, but other countries have also been raising their profile geopolitically and economically, including Mexico, Colombia, Chile, and Venezuela.

The often tense detachment is almost natural, like teenagers gradually coming of age and seeking their own destiny. Now it's up to the adult—the United States—to adapt to new circumstances in order to further its long-term interests.

Andrés Peñate has been dealing with geopolitics, oil, and policy for most of his life, both in government and in the private industry. He is a Colombian economist, a former international director and strategic advisor of British Petroleum (BP) in Venezuela, China, the United States, United Kingdom, and Colombia, and the former head of Colombia's now-defunct powerful intelligence agency, Departamento Administrativo de Seguridad, known for its Spanish acronym DAS. He was also vice-minister of defense under Uribe's administration and played a leading role in the country's successful strategy to decrease kidnappings and violent crime and to bring large swaths of Colombia back under the state's control. He lived through Chávez's ascent in Venezuela when working for BP, then through his consolidation as a close advisor for Uribe. Perhaps one of the most knowledgeable insiders in the complex relations between the United States, Colombia, Venezuela, and oil, he witnessed firsthand the rise of a populist left throughout the continent and Washington's incoherent responses. During an interview in Bogota, the Oxford-educated strategist reviewed the last few decades of US-Latin American relations.

> To say the US lost interest would be incorrect. It's a bit more subtle. Since the end of the Cold War the US abandoned Latin America as a commercial partner, and its role as leader and advocate of democracy was diluted. That said, the US didn't abandon the region altogether. It abandoned the strategy of achieving security through prosperity, and the agenda was taken hostage by issues of security and immigration.
>
> I don't think being part of the US agenda is necessarily a good thing anyway. It wasn't necessarily bad to be abandoned.

In other words, the detachment in the aftermath of the Cold War wasn't bad per se. It was the outdated, erratic, contradictory, and misguided policies that have undermined American interests since.

After the Soviet Union collapsed in 1989, relations with the region were downgraded. Bush "the first" implemented a realist policy that arguably sought new relations, but an end to his administration abruptly changed that. Under Bill Clinton the focus was antinarcotics war and free trade, and the United States failed to see the transition that was taking place in the region; last decade under Bush “the second” relations with the region soured as Latin America spread its wings and distanced itself from its old hegemonic overseer, and under Obama US policy has been more antagonistic than under his predecessors. At best, the United States remains far removed, if not indifferent, allowing idealist policies designed to placate domestic political groups to take hold.

To be sure, the problem is that American policy toward Latin America didn’t evolve. It got stuck. And meanwhile the regional transition has played out in many cases against American interests. If arguably the United States could afford to be shortsighted in the past, that is no longer the case. Chávez is the prime example, but policy trends in a bulwark ally like Colombia also point to more trouble ahead for the United States, especially in regard to energy.

Chávez’s appeal derives from his left-wing, populist promise of delivering an alternative to the neoliberal, unbalanced economic growth espoused by the United States. In exposing the flaws of the failed model, he created the hollow illusion of a bright future based on socialism. His alternative model is unsustainable in the long term of course, but that’s not what the United States should pay attention to. Rather, Washington should realize Chávez’s success does not lie in the solution he offers, but in harnessing the frustration of a failed US-driven economic model. The American recipe initially kicked off economic growth and a marked improvement in education, leading to the consolidation of a strong sophisticated middle class, but it fell short of addressing the severe social contrasts in Latin America. Indeed, Latin America has rejected Chávez’s pipedreams as much as Washington alternatives.

Chávez’s rise and the likely continuity of his revolution—at least in the short term if he doesn’t survive his cancer—must be dealt with in the context of what allowed him to rise to world stature in the first place. Confronting Chávez, and by extension the Bolivarian block, hasn’t worked, nor has ignoring him. Soft power, hawkish wars of words, and indifference have all failed to reverse gains made by anti-American forces. And this is an economic issue, not political.

But American energy and national security, like that of any country, is conditioned by public discourse. Venezuela is the best placed to improve American energy security and economic stability, but politics remains an obstacle.

Chávez has made undermining American policies a pillar of his policy and survival throughout his 13-year rule. Both countries are offering opposing visions of the future, but instead of prioritizing relations with the majority of Latin America, the United States instead decided to confront Chávez.

United States and Latin America: Imperialism, communism, and absence

US relations with Latin America have always been troubled. Up to the 1930s they were imperialistic. The Monroe Doctrine of 1823 made its hegemonic claim to the continent, warning Europeans to stay out. In 1904, Theodore Roosevelt issued the Roosevelt Corollary that basically restated European powers and threatened an American military response in case of an armed intervention in the continent, with the addendum that to preempt such an attack it was giving itself the right to do its own military intervening to stabilize countries. It did just that in Mexico, Dominican Republic, Cuba, Haiti, Honduras, Nicaragua, and Panama, to name a few, as part of its gunboat diplomacy.

It wasn't until 1928 that the United States started stepping back its imperialist policy, and in 1934 Franklin D. Roosevelt established the Good Neighbor Policy, renouncing military intervention, in an effort to shore up support ahead of World War II. Gunboat diplomacy though remained a strong deterrent against threatened US interests.

Twentieth-century US policy toward Latin America has its roots in the postwar creation of the Organization of American States in 1948. Most countries pledged to fight communism and that goal superseded all other interests, including democracy. The Cuban Revolution, led by Fidel Castro, only reaffirmed the communist threat was a greater one to the United States than the dictatorial governments that were sprouting up throughout Latin America. John F. Kennedy, in the two years before he was assassinated, was the only one to simultaneously try to prevent the rise of communism and antidemocratic totalitarian governments. He implemented the Alliance for Progress, an economic aid program meant to foster growth and better distribution of wealth to address the structural economic disparities that were fostering communist expansion in the first place, but it was vastly insufficient and conditioned to onerous terms. Lyndon B. Johnson and subsequent leaders gave up on the idea, instead supporting brutal military dictatorships in Brazil, Dominican

Republic, Chile, Paraguay, Bolivia, Uruguay, Argentina, Peru, Nicaragua, El Salvador, Honduras, Guatemala, Dominican Republic, Ecuador, and all Central American countries except Costa Rica, as long as they fought communism and protected American interests. In Central America, the United States also funded and armed ravaging civil wars to keep communism from expanding, especially after Nicaragua's revolution deposed the government in 1979 and imposed communism.

To be fair, US policies are by no means responsible for the gross inequality in Latin America or for the violent repressions. It was local elites in all cases, whether left or right winged, who implemented them. Armed struggles were funded by the Soviet Union and Cuba and later Nicaragua on one side, and by the US and traditional elites on the other. Latin America willingly hosted a proxy war.

The United States played its part, no doubt, by supporting the repressive regimes, not just with training, but with weapons and impunity. But the instability was ingrown and could in fact be interpreted as part of Latin American countries' natural evolution. The United States is not to blame for Latin America's chronic political and economic instability, a pervasive conclusion widespread in the region. Regional elites, mass illiteracy, and decades of discrimination and inequality are to blame.

Hundreds of books have been written about US hegemonic power up to the end of the Cold War and the debate is endless, complex and far from homogeneous, among other reasons because victims of Latin America's dictatorships and of economic oppression still make up a big part of Latin America's population. Each country has its own history and political and economic evolution. In any case, Cold War relations between the United States and Latin America are not an issue in this book. Suffice to say that the United States was anything but absent up to the Soviet collapse. Latin America was to an extent a priority, and the United States was heavily involved.

Modern relations characterized by US detachment began after the Cold War with George Bush "the first" in 1988, followed by Bill Clinton up to 2000, on to George W. Bush in 2008.

Francisco Santos, former Colombian Vice President under the Uribe administration and also one of the country's best-known journalists, summed up US absence during an interview in Bogota, Colombia:

> It's not recent. Latin America has been losing importance in the international sphere and the United States has not had a strategic

vision toward the region. Latin America weighed under George Bush senior, and a little before that, but not after. It's been about isolated issues, like the war on drugs. But the US vision of Latin America has been condescending and very wrong.

US detachment couldn't have come at a worse time. The 1980s were economically disastrous for Latin American countries, dubbed the "lost decade," and the region entered the 1990s in its worst economic shape in decades. Countries had borrowed heavily between 1975 and 1982, before the world recession, at annual rates of more than 20 percent. When recessions spread during the 1980s, several countries defaulted on their debt, starting with Mexico. Credit dried up and the region went into an unprecedented crisis. Income fell, inflation and interest rates soared, growth stagnated, and unemployment and poverty climbed. That is when the United States looked the other way.

The economic development model of import substitution industrialization was abandoned and countries adopted a new export oriented industrialization model. Globalizations advanced rapidly, especially after the fall of the Soviet Union. The new economic doctrine for the 1990s was based on the Washington Consensus that international financial institutions like the IMF and World Bank mandated, one that called for mass privatizations, tariff reductions, public spending cuts, and free trade.

Latin American economies returned to growth, but wealth distribution deteriorated. That is, the rich became a lot richer and poor a lot poorer. In fact, Latin America to this day remains the most unequal region of the world, meaning the gap between rich and poor is bigger than anywhere else.[5] It's not a coincidence that millions of poor in Andean nations like Colombia, Peru, Ecuador, and Bolivia resorted to the drug trade in the 1980s and 1990s to move up the ladder, as both white- and blue-collar jobs disappeared.

---

[5]Council on Foreign Relations—Independent Task Force Report, 2008, *US-Latin America Relations: A New Direction for a New Reality*: "All of these measures have had less of an effect on job creation and poverty alleviation than was initially indicated. The opening of Latin American economies, the successful correction of macroeconomic imbalances, and the reform of economic governance structures have not substantially lowered poverty or structural inequality. (Many argue that the fiscal and economic policies of the Washington Consensus have actually heightened inequality in the region, as the enacted policies did not address structural inequalities or microlevel bottlenecks.) These domestic political and institutional obstacles—in both Washington and Latin America—have had a corrosive effect on support for globalization and trade reform."

The economic crisis also contributed to the ultimate toppling of most of the brutal military regimes by the 1990s, especially as the United States began pressuring its old allies to give up power. Having defeated the Soviet Union and discredited communism, the United States cared only about nominal democracy and liberty in Latin America expressed through free elections.

Clinton and Bush "the second" reshuffled priorities and resources away from Latin America to consolidate America's newly found overarching global political and economic supremacy. The Middle East and Asia took priority, as did global economic growth. The 1990s left only a few positive headlines like NAFTA, the FTA signed by Canada, the United States, and Mexico, which came into force in 1994. The Clinton administration also convinced Latin American countries in 1994 to commit to the creation of the Free Trade Area of the Americas with the goal of creating a tariff-free continent from Alaska to the Patagonia by 2005. The rest was about combating resurging guerrilla movements, widespread corruption, weakening institutions, narcostates, and broadening instability.

Development aid and nation building largely dried up as the communist menace subsided. America, the father, then big brother, cared about little else other than the war on drugs, free trade, and stemming immigration flows for domestic electoral reasons. These policies had a compounded detrimental effect on millions of poor and middle-class Latin Americans, not on the increasingly slim wealthy class.

Free trade was especially disappointing. Mexico, for example, received millions in investment from American companies as of 1994. Between 1995 and 2000 growth averaged more than 5 percent annually, compared to 1.5 percent in 15 years prior. But corrupt elites mismanaged newly found wealth. The poor became ever so poorer while the wealthy amassed great fortunes that were not reinvested. In fact, the world's richest man, richer than any American, is Mexican. But policymakers were too busy to notice how their detachment would backfire. Despite NAFTA and billions in foreign investment, poor Mexican immigrants flooded the United States, drug cartels prospered, and the government grew increasingly unable to foster stability. US-imposed development models were creating wealth, but also grassroot instability.

In the meantime, several elected Latin American leaders ransacked their countries and corruption undermined any prospect for change. Open borders still attracted billions in investment, but most of the population didn't prosper and instead endured a gradual decrease of their real incomes. The United States was oblivious, or indifferent, to growing instability in the region, as long as people voted and governments cooperated with its highly unpopular priorities. As long as there were free elections and markets were open, the rest would solve itself, a distant Washington concluded. But it didn't. It got worse.

The United States unknowingly pushed a political and economic transformation that set the stage for the rise of the Latin American left, one that rejected US hegemony and leadership. Masses were disenchanted with America's vision of the future and the ground was ripe for yet another radical shift toward the left, one led by Venezuela's Hugo Chávez, who captured the presidency in 1999 on the promise of deep-rooted economic and political reforms.

Like all other American presidents, George W. Bush came to power in 2000 with the promise of swift policy reforms toward Latin America, of reversing a decade of absence, of a new push to turn back a tide of growing lack of governability at the hand of the drug trade, corruption, and rising populism.

But two events undermined US-Latin American relations: the September 11, 2001 terrorist attacks and Bush's tacit support of a military coup in Venezuela to depose Chávez. To be fair, everything changed after the Al Qaida attacks on the United States. American muscle and attention turned to fighting terrorism and the war on Iraq. Latin America fell even further out of the map, but so to a lesser degree did a rising China and Russia. Even Iran.

But the Bush administration's perceived support in 2002 of a military coup in Venezuela forever changed relations with an outraged region. It's not that other leaders wanted Chávez in power, far from it, but it was an unforgivable reversal of the democratic principles that the United States had finally adopted in the early 1990s. The very suggestion that Washington would once again support military coups put the entire region on the defensive, starting with powerhouse Brazil. At stake was not Venezuela's democracy, but setting a precedent for a renaissance of the century of dictatorships.

The new Latin American left-of-center regimes in Brazil, Argentina, Uruguay, Bolivia, Chile, Nicaragua, Ecuador, and eventually Paraguay and El Salvador—that is, most of the continent's population living precisely in countries that suffered at the hands of American-backed militaries—grew suspicious of Washington's intentions and its democratic U-turn.

Not only were the Bush administration's policies defeated hand-me-downs that were backfiring, but they were dangerously flirting with military alternatives to contain a perceived threat from a rising left. The fallout spread throughout the region, with a few exceptions like Colombia, which was receiving billions in military aid.

Economically, the region was delivering macroeconomic stability modeled on the Brazilian and Chilean experience, after directly rejecting the continental FTA endorsed by the United States. The drug war was being lost on all fronts. Worse, it moved next door to the United States. And immigration clampdowns only enraged a newly empowered Latin America, without really denting inflows. US policies had backfired for both sides, and still Washington's leadership was nowhere to be found.

Had all countries chosen the Brazilian or Chilean models, ultimately pro-market and pro-reform, while addressing unbalanced wealth distribution, American interests would not have been threatened. In fact, these models are bringing about the long sought goal of economic and political stability and the creation of a big middle class eager to consume American products.

But a block of countries opted for Chávez's model, the Bolivarian movement, all united under an anti-American, statist banner. Chávez went further than any other, and added a militaristic component.[6] And they had oil to further their interests. US confrontational policies no doubt contributed to the consolidation of Chavismo, if not to its rise. And even if Chávez has failed to deliver on a viable alternative to America's vision, he has succeeded in undermining its clout significantly, to the benefit of Brazil.

In 2008, a Council of Foreign Relations taskforce of academics and experts on Latin America drafted some policy recommendations. The conclusions described the new reality.

> For over 150 years, the Monroe Doctrine provided the guiding principles for U.S. policy toward Latin America, asserting U.S. primacy in the foreign affairs of the region. Over the past two decades, those

---

[6]Javier Corrales and Michael Penfold, *Dragon in the Tropics: Hugo Chávez and the Political Economy of Revolution*, Washington, D.C.: Brookings Institution Press, 2011, p. 3: "Over the past decade many of the region's leaders have deployed some of these practices, but none has undermined check and balances institutions and co-governed with the military to the extent that Chávez has in Venezuela."

> principles have become increasingly obsolete. Washington's basic policy framework, however, has not changed sufficiently to reflect the new reality. U.S. policy can no longer be based on the assumption that the United States is the most important outside actor in Latin America. If there was an era of U.S. hegemony in Latin America, it is over.
>
> U.S. policymakers must change the way they think about the region. Latin America is not Washington's to lose; nor is it Washington's to save. Latin America's fate is largely in Latin America's hands. A failure to acknowledge how Latin Americans define their own challenges has created new political strains in recent years. It has also caused U.S. policymakers to overlook the ways in which the United States can meaningfully contribute to Latin America's progress—furthering the United States' own interests in the process. By truly beginning to engage Latin America on its own terms, Washington can mark the start of a new era in U.S.-Latin America relations.
>
> Opening economies, strengthening democracies, and fighting drug production and trafficking remain important priorities. But continuing to build U.S. policy on these pillars alone reflects a mistaken sense of what U.S. policy can realistically achieve and a failure to recognize where Washington can meaningfully bolster Latin Americans' efforts to improve their own quality of life, providing a new foundation for U.S.-Latin America relations in the process.

Latin America was simply no longer America's backyard when this century began. The United States just failed to notice.

Chavismo, Bolivarianism, and Oil

Hugo Chávez, his rise, and his ability to undermine American interests are all related to oil. Chávez would not be who he is without it. Nor would the United States treat a regime like his the way it does if he didn't control so much current and future oil production. Oil is simultaneously his lifeline to remain in power and to expand his movement regionally in the short term, and the only means to contain him in the long term.

Chávez came to power following several decades of mismanagement and corruption at the hand of Venezuela's traditional political parties that saw poverty and the gap between rich and poor increase, despite the country's oil wealth.

Former Colombian Vice President Santos on Chávez's rise:

> Chávez's phenomenon is internal. I don't think the US is responsible. It's institutional to Venezuela, of extreme corruption of the traditional parties that allowed a cure to surface that was worse than the sickness.

Chávez promised to reverse Venezuela's history of inequality. Javier Corrales, who teaches political science in Amherst, and Michael Penfold, a political economy professor in Venezuela, in 2011 addressed Chávez's rise in their book *Dragon in the Tropics*:

> Some scholars have argued that Venezuela's legendary democracy—one of the first successful "pacted" transition in Latin America—turned into a rigid "partyarchy" in the 1980s. During this period two parties dominated the political field. Far too many actors across all income categories were excluded by an agreement among these parties' leaders that over time restricted access to democratic institutions and failed to manage economic development once oil fiscal resources started to decline. This led to demands for new and more participatory political institutions, a tide that brought 'revolutionary' Chávez to the fore. Others have argued instead that Venezuela's experiment with market economic reforms in the 1990s led to harsh austerity policies that expanded poverty without restoring growth, leading to a demand for a more leftist-populist-nationalist type of economic development. Finally, a new wave of high oil prices in the early 2000s supplied the means for the Chávez regime to deliver on these society-demand changes. In a nutshell, many scholars think that the failure of liberal democracy and economics explains the demand for the type of regime now in place, while oil provided the figurative and concrete fuel.
>
> Venezuela's economic ailments resulted from factors other than "neoliberalism," namely, the persistence of dependence of oil, which caused macroeconomic volatility; political party fragmentation, which triggered policy incoherence and infighting; government mismanagement of the economy, which led to greater contraction of the private sector in the 1990s; and the Asian crisis of 1997, which devastated Venezuela's economy just around the time that Chávez ran for president.

In any case, Chávez has built himself into one of the world's most recognizable—and polarizing—figures. His persona transcends continents, languages, and cultures, and is probably evenly split among admirers and hecklers. There is Chávez the liberator, Chávez the clown, Chávez the revolutionary, Chávez the buffoonish martyr, Chávez the dictator, to mention just a few of his guises.

But to gauge the threat of Chávez to US interests, he must be understood as a populist, autocratic politician. To label him ideologically is both superficial and incorrect. Politicians like Chávez, regardless of vocal antagonism toward the United States, are driven by their survival instinct, not ideology. Chávez's brand of politics has been dissected for decades in the myriad of developing country left- and right-wing *caudillos*.

Populists are above all pragmatists, which explains why they can fit into any part of the ideological spectrum, guising behind co-opted labor movements that crown self-proclaimed left-wing revolutionaries, or populist nationalists like Vladimir Putin of Russia and former Colombian President Uribe. Africa has had a long history of such leaders that have flip-flopped from left to right in a whiplash. The simple goal of populists, unlike ideologically driven parties, is to retain power, if not for themselves, for their ego-driven movements. Populism is antidemocratic by definition, but rendered politically stable by increasing economic devolution to the masses, especially through wasteful subsidizing, scapegoating a common enemy, propping the economic interests of a new ruling elite, and gradual takeover of a state's institutions. And it's often accompanied by a powerful military that manipulates nationalism to covertly hang on to power.

The closest match to Chávez in fact is not any left-wing ideologue, despite his self-built image as the heir of Fidel Castro of Cuba. It's Juan Domingo Perón of Argentina, a right-leaning populist who used his vast support among the labor movement to be elected three times. He supported military coups and was deposed by one himself. He was also jailed, forced into exile, and still he managed to return to power, until he died in his third term. Perón used mass populist support of the left, along with radical right sectors of the Armed Forces and the Catholic Church. Any deal was possible, as long as he remained in power. His fragile coalition included communist, socialist, land owners, capitalist, and students. In practice, he was more a fascist and his sympathies for the Nazi regime are documented. In his second term he was deposed by a mélange of socialist, right-wing Catholic supporters, and conservatives, many of whom had previously been allied with him.

Colombia's former intelligence chief Peñate on the similarities between Chavismo and Peronismo:

> Much of what Chávez does is caudillo populism. From a historical point of view, Chávez is Perón. It's not statization of the left, but Caribbean Peronismo. What allows him though to flaunt his left-wing credentials is Cuba, anti-Americanism, and ties to revolutionary movements. It's very useful with the radical left in Europe. In the US the left might be outcasts, but in Europe they are mainstream.

Chávez too is a charismatic leader of a coalition of both right- and left-wing movements, and he is able to hold on to power with the militant support of most poor Venezuelan and the Armed Forces. The military is checked by the masses and the masses are checked by the military, in a textbook divide and conquer strategy that allowed the likes of Adolf Hitler, Benito Mussolini, and Perón to hold on to power. That is not to say that Chávez is fascist, although part of his coalition does strongly resemble it. Chávez leans more toward the left, no doubt, but he is by no means the dogmatic Marxist ideologue that Castro is. His coalition is also fragile and includes a majority of the political spectrum.

Chávez, a paratrooper, rose to stardom in 1992 when he led a failed military coup as colonel. He was jailed and then released on an amnesty. But he embodied mass discontent with traditional parties that had governed for decades. Poverty almost doubled in the decade between 1984 and 1995 to more than 66 percent. Inflation between 1993 and 1997 increased annually between 50 and 60 percent, peaking at nearly 100 percent in 1996. By December 1998, when Chávez won the presidential election with 56 percent of the vote, the per capita GDP was at the same level as the early 1960s, and the purchasing power was equivalent to a third of what it was in 1978.

In 1999, Chávez ended years of pro-American oil policy which disregarding OPEC oil quotas that, until then, were allowing the market to keep well supplied. He also concentrated power around himself by pushing constitutional reform, then gradually nationalizing private institutions.

In effect, Chávez replaced the existing corrupt elite with his own corrupt followers. Another regime moved to squander the country's oil wealth. And to hold on to power, he legally purged dissent, starting with the Armed Forces and then Petróleos de Venezuela SA, best known for its acronym PDVSA, the country's breadbasket. He was reelected in 2000 under the new constitution with almost 60 percent of the vote, and in 2001 he allied himself with OPEC hawks like Iran and Iraq to drive oil prices upwards.

Venezuela's opposition meanwhile had no way of countering Chávez. They harshly criticized his model, but offered little alternative other than a return to the vastly unpopular status quo of yore. In 2001 Chávez began a land reform to seize lands from traditional elites for redistribution among poor peasants. By the end of the year, the country's business leaders called for a strike. At the beginning of 2002, army officers and PDVSA management started demanding his resignation, backed by the Catholic Church. Chávez answered by firing PDVSA managers and mobilizing his supporters.

On April 11, 2002, a massive opposition march clashed with Chávez supporters in giant riots that also pitted the National Guard against police. A group of generals and admirals went on television saying they no longer recognized Chávez's authority because of his inability to contain violence that killed 11 people. The opposition said Chávez supporters were shooting, while Chávez said the opposition had orchestrated the riots and instability. At midnight, Chávez had allegedly agreed to resign after coup leaders threatened to attack the presidential palace surrounded by this supporters. Before sunrise Chávez had surrendered and was taken to a military base. Business leader Pedro Carmona later on April 12 was sworn in as interim president and his first action was to revoke the constitution, civil liberties, freedom of press, and so on. Most Latin American leaders condemned the coup immediately, with the notable exception of Colombia.

Throngs of Chávez supporters mobilized and brought Venezuela to a standstill. Chávez sent a message out to his supporters denying he had ever resigned and that he was being held hostage. Army loyalist eventually managed to free him, by which time coup leaders were on the run.

But before that could happen, the Bush administration rushed to recognize Carmona, pledging to work with his interim government, which was in the process of denying all civil rights. Here is the statement issued by the US embassy[7] in Caracas then:

> In recent days, we expressed our hopes that all parties in Venezuela, but especially the Chávez administration, would act with restraint and

[7]www.state.gov/documents/organization/153060.pdf

> show full respect for the peaceful expression of political opinion. We are saddened at the loss of life. We wish to express our solidarity with the Venezuelan people and look forward to working with all democratic forces in Venezuela to ensure the full exercise of democratic rights. The Venezuelan military commendably refused to fire on peaceful demonstrators, and the media valiantly kept the Venezuelan public informed.
>
> Yesterday's events in Venezuela resulted in a transitional government until new elections can be held. Though details are still unclear, undemocratic actions committed or encouraged by the Chávez administration provoked yesterday's crisis in Venezuela. Venezuelan military and police refused orders to fire on peaceful demonstrators and refused to support the government's role in such human rights violations. The government prevented five independent television stations from reporting on events. The results of these provocations are: Chávez resigned the presidency. Before resigning, he dismissed the Vice President and the Cabinet. A transition civilian government has promised early elections.

In private, the US Assistant Secretary of State for Western Hemisphere Affairs Otto Reich summoned continental ambassadors to reportedly back the new Venezuelan leadership and pleaded with others for similar support, which he didn't get. American ambassador to the Organization of American States (OAS) at the time Roger Noriega recognized a "provisional government."[8]

The White House also admitted its diplomats had met several times with coup leaders before it actually took place, but it denied endorsing it. Ultimately, an internal government investigation determined no officials were responsible for wrongdoing, for promoting a coup. That said, officials did little to impede it. The United States did not condemn the coup until after Chávez had been reinstated to power.

Luis E. Giusti, the former chairman and chief executive officer of PDVSA, fled Venezuela on the very day Chávez was sworn in. He has lived in a self-imposed exile and is currently a senior advisor of the CSIS, a board member in numerous organizations and companies, and a regular speaker on global energy issues. He also happens to be one of

---

[8]www.state.gov/documents/organization/153060.pdf

the foremost experts of the American continent's oil sector and a savvy OPEC insider.

He led one of PDVSA's most profound transformations known as La Apertura—the Aperture/Opening—attracting private investment of some $30 billion during the ten years that followed his rise to the stewardship of the company in 1995 after decades working as a petroleum engineer. He was also the last CEO before firebrand Chávez came to power in 1999.

Like many Venezuelans, he disputes US involvement in the coup, as well as the sequence of events that took place that day.

> The evolution of the coup in Venezuela has been for long time misunderstood, or more accurately distorted. People marched peacefully in the streets of Caracas. Chávez ordered the activation of Plan Avila, which essentially meant that tanks take to the streets against the march. The Top Brass disobeyed the order, which eventually forced the military to take over. No coup so far. The Top Brass call Carmona to assume transitory power, but the man starts making a mess by dissolving all institutions, Congress, Governors, mayors, regional legislature, and many others. In addition he practically swears himself as president and announces a cabinet totally disconnected with the crisis. The coup becomes effective. Facing the mess created by Carmona, the military take over again and decide to bring Chávez back. It is false that some military deposed him and some other loyal to him rescued him and brought him back. It was all a single process, and that is why except for a couple of retirements, never any military went to jail.

Regardless, the coup changed the history of Latin America. It emboldened Chávez in his anti-American rhetoric and made the rest of the continent suspicious of American intentions. The United States and Colombia, with the initial support of Spain, were outcasts in the region for their early tacit support of the coup.

Internationally, the antagonism of the Bush administration and Colombia's Uribe also consolidated Chávez's hold on power. As a populist, he created a persecution ethos in Venezuela. US imperialists, using Colombia as a launching pad, were out to depose him, to reverse his revolution, Chávez claimed, and his supporters took a militant defense of their leader.

The opposition still had one last card. A second general strike, this time including PDVSA, started in December 2002, and more military dissenters surfaced. PDVSA came to a standstill for two months. But again, the opposition crumbled and gave Chávez more ammunition to blame corrupt Venezuelans helped by American imperialists that were out to get him. Indeed, the coup and subsequent strike allowed him to accelerate the purging and tightening of his hold on power, especially in the Armed Forces and PDVSA. He was reelected in 2006, the same year he nationalized oil fields managed by foreign companies, increasing the government share from 40 percent to 60 percent.

Chávez's power is magnified by the lack of alternatives, despite the deteriorating effect on the country's economy. Michael Shifter, president at the Inter-American Dialogue, a policy think tank, and adjunct professor of Latin American Studies at Georgetown University's School of Foreign Service, describes the opposition in Venezuela:

> The opposition's lack of success stems from its past unwillingness even to recognize—let alone devise solutions to—the deep social problems that Chávez has identified. Chávez's government, meanwhile, has undertaken important social programs and launched workers' cooperatives in urban slums. Venezuela's oil wealth has made massive expenditures possible which have undeniably had some effect.[9]

Some experts describe Chávez government as a hybrid regime in which freedom exists, but the system of checks and balances is broken.[10] Internally, any person or group can criticize freely, from press to political parties, but at the risk of becoming easy prey to his rhetoric. As a populist, it's either for or against him.

Chávez also quickly realized that his ability to retain power would require exporting his revolution to other countries. This involved legitimate support for left-leaning governments aligned with his objectives, and covert, illegitimate support for proxies that supported his Bolivarian movement, one that resurrected the dream of Simon Bolivar, the nineteenth-century military and political leader who was instrumental in winning the independence from Spain of most South American countries.

---

[9]Shifter. In Search of Hugo Chávez. Foreign Policy.

[10]Corrales and Penfold, *Dragon in the Tropics*, p. 1.

Peñate, the former BP strategic analyst and former head of Colombian intelligence services, analyzes Chávez's rise to regional leadership:

> I know many in the US think Chávez is filling the space that the US left behind. But I don't think a more involved US would have prevented it anyway. Chávez has exploited the inability of Latin American countries to create wealth. And that is not the fault of the US. We've been independent for over 200 years!
>
> Chávez has been able to capitalize on the discontent in our region because the success of capitalism hasn't trickled down to the masses. It's not America's fault, but our own. But the US hasn't played a very illuminating role in this sense either because it only cares about its security and immigration agenda.
>
> The US grew rich thanks to democracy, immigration, and free trade, but to its Latin partners it doesn't offer free trade, democracy, or labor mobility.
>
> The reasons why it's cheaper for a peasant to grow coca leaves instead of potatoes is that cocaine does have its own free trade agreement. How can a cow in Belgium get more money in the form of subsidies than an entire peasant family of the Latin America? Countries became big drug producers not because they are wired to delinquency but because most peasants didn't have a choice.
>
> This is the discontent that Chávez exploits. That is how he grows.

In 2004 Venezuela and Cuba created the ALBA group of nations, which stands for Alianza Bolivariana para los Pueblos de Nuestra América, as an alternative to the US-sponsored continental FTA. Bolivia joined in 2006, Nicaragua in 2007, Honduras in 2008, Ecuador in 2009, as well as several small Caribbean islands along the way. Other countries, while not joining ALBA, cooperate closely with the group, like Argentina, Uruguay, and Paraguay. The allure was cheap oil or credit with very favorable terms.

Venezuela supplies its top regional ally Cuba, for example, with about three-fourth of its oil since 2000, today equivalent to around 120,000 barrels per day (bpd). Cuba pays for this oil, worth at least $3 billion a year depending on prices, with doctors, dentists, and human capital of all sorts, including intelligence services. Under the Petrocaribe initiative, Venezuela also provides a big chunk of oil and refined products. It finances half of the shipments for as long as 25 years at below market rates, but in return it becomes the biggest benefactor of receiving countries, giving it considerable clout. In all, around 400,000 bpd of oil are shipped to Caribbean and Central American countries, according to the EIA. Chávez has in fact subsidized oil for heating for some 400,000 poor Americans in 25 states for 7 years, at a cost of about $400 million.[11]

Chávez's alternative is based on 21st Century Socialism, a nationalistic ideology that seeks strong state control mixed with conditioned private enterprise, economic self-sufficiency, participatory democracy, equitable distribution of natural resources, and other headline idealist objectives. Chávez has acknowledged countries need to transition toward socialism, an excuse of sorts he has used in the past to justify his blunders and authoritarian regime.

Venezuela has also strongly pushed for more regional integration around ALBA institutions, but so far without much success. Countries like Argentina share some objectives, but vary on how to achieve them. Most other countries fit into some sort of resurgent, pro-market socio-democratic spectrum from the Brazilian center-left model to Colombia's center-right. Even Colombia and Mexico, the bulwark ideological US allies, have moved more to the center, and all have resorted to lesser degrees of nationalism and populism.

During an interview in Bogota, former Colombian President Uribe said oil, not Chávez's left-wing message, explains the success of Bolivarianism.

> There is an issue of rhetoric, but not reality. The reality is oil has allowed him to advance the Venezuelan process, which has many traits of a dictatorship. He has been able to advance not because of his rhetoric but because he has had a lot of oil to gain popular support.

Oil, no doubt, has underwritten Chávez's foreign policy. It's difficult to calculate exactly how much Venezuela has contributed to Latin American development programs due to the lack of transparency and the big gap between promise and reality. One estimate published by the Latin American Social Sciences Institute, an intergovernmental

[11]Refers to the CITGO-Venezuela Heating Oil Program: http://venezuela-us.org/2011/12/13/in-its-seventh-year-citgo-venezuela-heating-oil-program-helps-400000-needy-americans/

autonomous organization, estimated that between 1999 and 2009, Venezuelan international cooperation commitments were nearly $40 billion, more than four times higher than US foreign aid contributions.[12] The number can be misleading because most of Venezuela's aid is in the form of energy infrastructure investments made by PDVSA that would have been hard to finance otherwise. But whatever the numbers, there is consensus that Venezuela has pumped Latin America with significantly more development aid than the United States since Chávez has been in power, and that in itself is very telling considering the US economy is exponentially bigger than that of Venezuela.

Peñate on Venezuela's oil diplomacy:

> A big part of Latin American countries are oil importers and some are very poor countries importing oil. When oil prices go up, the option to get cheaper oil is a political element similar to one the Soviet Union had, but to a smaller degree. That is how Venezuela got influence over Cuba. I know most people think Cuba is the one influencing Venezuela, but it's the other way around. Subsidized oil is what gives Chávez oxygen. And he does this all over the place, even in the US.

But Chávez's most important leverage is inside Venezuela. In his 13 years in power, vast economic reforms, from agriculture to healthcare and education, have reduced poverty more than any other country in Latin America, aside from Argentina, which was recovering from its worst economic crisis years prior.[13] Moreover, income distribution in Venezuela improved more than in any other Latin American country since 2002. The Venezuelan government claimed on January 2012 that it has invested $468 billion into social spending, including 60 percent of its budget in 2011,[14] since Chávez came to power in 1999.

But Chávez's economic model is based on an unsustainable use of oil revenue. It's self-defeating in the long term and will soon require a major

---

[12]Venezuela: Política exterior y rentismo, published by FLACSO: www.flacsoandes.org/internacional/publI_acade/venezuela/05romero_y_curiel.pdf. And US foreign aid data from the State Department, Congressional reviews, and USAID disclosures.

[13]2011 ECLAC (United Nations) Social Panorama of Latin America report. CEPAL report Panorama Social de América Latina.

[14]Venezuelan embassy in the United States: www.venezuela-us.org/live/wp-content/uploads/2009/08/Fact-Sheet-13-in-13.pdf

pro-market reform. Shifter, president at the Inter-American Dialogue, describes the limits to Chavismo:

> Literacy programs have been afforded high priority and have made some progress. Regardless of whether the conditions of Venezuela's poor have marginally improved or marginally worsened under Chávez, his "Bolivarian Revolution" is hardly a sustainable model for Venezuela's or the region's predicament. Its approach is fundamentally clientelistic, perpetuating dependence on state patronage rather than promoting broad-based development. Random land-reform measures and occasional confiscations of private property have had less of an economic than a political and symbolic rationale. Crime, a dominant concern for Venezuelans, has gotten worse.[15]

Alejandro Grisanti, a senior economist for Venezuela, Ecuador, Peru, Central America, and the Caribbean at Barclays Capital Inc., also analyzed Chavismo's economic failings in an article published in the Spring 2011 issue of *Americas Quarterly*:

> Since 1998, and despite the oil windfall, Venezuela has grown on average 2.3 percent per year. In contrast, the seven largest Latin American countries grew on average by 3.2 percent. Within this period, Venezuela experienced five years of negative growth with an average contraction of 5.6 percent and seven years of positive growth with an average expansion of 8.4 percent. Venezuela is an extreme example of Dutch Disease: in nominal terms, oil exports have grown 410 percent since 1998, while imports have grown 130 percent. But in real terms, exports have fallen 40 percent, with imports more than doubling.
>
> At the same time, Venezuela's institutional development lags behind its Latin American peers. Using the World Bank Governance Indicators for 2009, Venezuela's percentile rank for control of corruption is 8 percent; the regional average is 47 percent. Similar selective comparisons are as follows: rule of law (3 percent Venezuela, vs. 43 percent regional); quality of regulation (4 percent vs. 45 percent); government

[15]In Search of Hugo Chávez: Foreign Policy, May 2006: www.foreignaffairs.com/articles/61703/michael-shifter/in-search-of-hugo-ch%C3%83%C2%A1vez

> effectiveness (19 percent vs. 50 percent); political stability (11 percent vs. 38 percent); and accountability (26 percent vs. 49 percent).
>
> A country like Venezuela, with abundant natural resources, should be seen as having high potential investment in various projects and abundant access to foreign exchange to purchase capital goods and raw material. But unfortunately, after 12 years of unproductive economic policy, Venezuela has spent more than $517 billion in oil income and increased its total debt from $31.3 billion in 1998 to an estimated $96 billion at the end of 2010.

That doesn't include at least $30 billion China has loaned to Venezuela through several agreements, to be paid for with oil shipments currently equivalent to at least 400,000 bpd.

To be sure, oil has allowed Chávez to consolidate and expand his revolution, but the real potential has been squandered. The Chávez model might be a short-term alternative to the neoliberal model of the 1990s championed by the United States, but it's clearly not a better model than the socio-democratic models of Brazil and Chile.

All South American countries have to some degree distanced themselves from US leadership, forging instead regional integration, one led by Brazil, which also serves to contain Chávez's more radical alternative. It's equally untrue to understand this Latin American phenomenon as one designed by Chávez. If anything, Brazil has remained the discreet leader, positioning itself at the center of the other center-right and center-left models, with Colombia and Argentina now on opposite sides.

Shifter describes the Chavismo phenomenon in a continental context:

> From the outset, it has been clear that Venezuela, with a population of 26 million, is too small a stage for Chávez's ambitions. Chávez has taken full advantage of a confluence of favorable factors—lots of money, Latin America's political disarray, U.S. disengagement from the region, widespread hostility to the Bush administration—to construct alliances throughout the Western Hemisphere and beyond. He has skillfully managed to establish himself as a global and regional leader, using oil money and brash anti-Americanism to attempt to construct a counterweight to U.S. power.
>
> But even as Chávez can help shape a regional environment favorable to such populist politicians, it is inaccurate to blame the rise of left-wing

> candidates on his influence or his scheming. Those figures are products of particular circumstances, and they would be contenders without Chávez. Chávez's shrewd use of resources, calls for social justice, and fierce attacks on an unpopular U.S. administration have had such resonance in Latin America precisely because leaders such as Evo Morales of Bolivia are reacting to many of the same frustrations that gave rise to Chávez in Venezuela. Chávez's adherents include not only remnants of the region's unreconstructed left, but also many who are simply frustrated by failed economic and political models and are searching for answers.
>
> Accordingly, many Latin American leaders, such as Lula and Kirchner, indulge him and accept his attractive economic deals without really endorsing his agenda.

Chávez, on the other hand, is increasingly hard pressed to attract more allies to his ALBA-led organization, to Bolivarianism. This is the result of the disastrous economic effects of his policies, which even Ecuador is reluctant to implement. But no doubt, his groups still holds significant power and influence in terms of how South American, or Latin American unity evolves. His block is strategically placed in Central America, South America, and the Caribbean, and his oil windfall, even if wasted, allows him to project enough power, at times even rivaling Brazil.

That external projection though also shields his troubles back home. Chávez unquestionably is Venezuela's most popular leader, by far. He has constantly received positive approval ratings from the majority of Venezuelans, as well as a comfortable lead in voter preference.

Chávez's populist success is based on his charismatic leadership, one able to hold together a wide spectrum of political groupings. In 2007, he was able to unite his coalition under one same party, further concentrating power around him. The United Socialist Party of Venezuela is a heterogeneous mesh of Chávez's Fifth Republic Movement and smaller parties, including nationalist, socialists, farmers, unionists, but not communists. A small part of his coalition partners in fact decided against joining the new party.

The 2010 parliamentary elections, the first in which the opposition decided to partake, still gave Chávez's party 48 percent of the vote and 98 of 165 seats in parliament, despite a two-year recession. The coalition of opposition parties got 47 percent of the vote and 65 seats. The other seats went to a smaller center-left party close to Chávez's movement.

In February 2012, Henrique Capriles won the primaries of Venezuela's fractured opposition. It's the most cohesive political effort to challenge Chávez so far. Capriles is a long-shot candidate, but the opposition's cohesion around him illustrates that Venezuela is facing a decade of reform, with or without Chávez.

Indeed, Chávez's political threat is not the opposition, but his own coalition. His power depends on his ability to keep it together. The coalition is cemented and enforced through his hold on security and intelligence services, which itself is also a shaky mix of nationalistic, Marxist, conservative, and socialist officers, often at odds with each other. That is further complicated by the visible and influential Cuban intelligence services, which ultimately check the entire apparatus.

Peñate, who lived through the political rise of Chávez from coup leader to regional leader, analyzes internal divisions within Chávez's coalition.

> People don't understand Chávez's government is a coalition. It is a fusion of people who shouldn't be together. A part of Chavismo is right-wing, made up of military coup allies. Another part is radical left, and the majority of people are opportunistic.
>
> If you take away Chávez's rhetoric, there is very little of a radical left in him. All his ministers are rich. They might put on a T-shirt of Che Guevara, but they have Cartiers on their wrists and US visas to go shopping in Miami like they used to do before. These are no Castro-likes.

But ultimately Chávez's legitimacy, his resilience and ability to perpetuate himself in power, comes from free and transparent elections certified by numerous international organizations, including the Carter Center. Like other populist leaders, he has manipulated constitutions to further his policies, but he has played by the very rules that the United States set as benchmarks. His constitutional reforms have been approved through several referendums. He also survived a 2004 recall referendum to revoke his mandate, and has been elected president by a majority three times in 12 years.

Again, here's Peñate on Chávez's use of democracy:

> Chávez tried a coup and failed. He came to power democratically and has stayed there via democratic rules longer than anybody else in this continent. He understood that as long as he kept the formality of a

> transparent vote, he could undermine democracy. As long as he takes a majority of public opinion into account, he can't be accused of being a dictator.
>
> Indeed, Chávez understood being a dictator is not a good business, that it's better to be sensitive to public opinion, to buy public opinion, and to win elections. He has become an expert in winning elections without fraud.

The strategy is effective and has spread to other Latin American nations in varying degrees. Populist ALBA leaders have also pushed through massive constitutional reforms that on the one hand favor the economic inclusion of impoverished masses, but on the other hand guise authoritarian powers by increasing the executive's control over the State, an ultimately antidemocratic result achieved through democratic means. They include Rafael Correa of Ecuador, Evo Morales of Bolivia, and Daniel Ortega of Nicaragua. But even Uribe of Colombia used his overwhelming popularity to push through legislation to perpetuate his power and to attack the independent judicial branch that was checking his intentions. Neopopulist leaders in the left and right have attempted, and succeeded in many cases, in translating political popularity into broadened powers for themselves, including Argentina, Uruguay, Paraguay, Peru, Costa Rica, and others.

Indeed, it's the malice of Latin America's democratic process, one that was ironically imposed by the United States after the Soviet collapse. For so many decades transparent vote casting was the only goal. An electoral victory was practically a license to misgovern in most of the region, regardless of ideology. For years, Peru and Ecuador have been plagued by corrupt megalomaniacs that were nonetheless elected democratically.

That is how the Obama administration inherited US-Latin American relations, an issue we'll pick up again in Chapter 8. Like others in the past, he promised new relations and new ideas. But again, the reality is different. US policy has changed little and leadership is nowhere to be found.

3 GAUGING CHÁVEZ AS A THREAT

As we analyzed in previous chapters, America's regional policy is ultimately driven by domestic politics. In Venezuela's case, how much oil the country produces is also an equally powerful issue. Whatever the rhetoric, Chávez's oil immunity is simply unavoidable.

Oil wealth in the context of geopolitics is intrinsically tied to American policy, whether it's Saudi Arabia, Iran, Russia, or Venezuela. That doesn't mean oil is the main driver in foreign relations or that the United States uses its military might to control oil production, a cynical and simplistic conclusion. But many oil producers and consumers, and certainly the biggest ones, use oil as a deterrent to some degree or another. For producers, it's an economic shield that weighs heavily on policy-making, even if consumers like the United States prefer not to admit it.

There's also a long and diverse history of using oil as a weapon, starting with the Arab oil embargo. And more recently, the West is also targeting oil income, in the case of Iran, defying even long-held unwritten rules that the free-flow of crude should never be politically manipulated. In practice, the oil threat is only effective if producers can't export their oil, or if consumer economies can't afford sustained high oil prices.

It's also a double-edged sword. The oil embargo that the United States and Europe imposed on Iran has been painful, but not nearly enough. Iran can still export enough oil, at least for now. Meanwhile, Western countries, especially Europe, are suffering the consequences of higher oil prices. The ultimate winner remains to be seen.

Iran and Venezuela often also threaten to cut oil flows, but producer countries stand to lose more in these cases because of their dependence on oil exports. To put it another way, the United States and other consumers might suffer from higher prices, but producers suffer more in the long run. Still, using oil production as a shield can be very effective. For example, the US ability to threaten Iran militarily over its nuclear program is conditioned—not determined—by the impact on oil markets.

At this juncture though, oil producers know that the supply and demand balance will remain tight throughout this decade, meaning oil prices will be more vulnerable to any threat of long-term disruption. American involvement in Libya was also conditioned by oil prices. The IEA in July 2011 warned that a prolonged conflict would undermine its emergency oil relief effort and even OPEC's ability to offset lost Libyan production.[1] Libyan oil production has recovered much quicker than expected and it will likely reach prewar highs sometime in 2012, but the political stability required to increase output beyond that is still uncertain, in no small part because rival militias are still battling it out for control of the country.

The new threat to global supply is the standoff between Iran and Western countries, including the European embargo that will go into full force in July 2012. Sudanese and Algerian production is also under political risk of underperforming, not to mention Venezuela's. Saudi Arabia alone cannot offset compounded supply losses, despite pumping at a 30-year record. But other OPEC countries have not been able to boost output as much and as a result markets are stretched thin, despite the economic slump. A long-term shut-in of oil production in Iran or Venezuela, not to mention of even a small part of Russia's, would be challenging to offset.

The oil shield alters all calculation. The more the United States pressures Chávez, the more Venezuela's oil production falls and oil prices rise. The same could be said of Ecuador, even if its production is minimal. That said, oil does not make a country invincible. In a case involving direct security threats like that of Iran, oil price concerns are trumped. But Venezuela is a different story.

Is Chávez a threat?

American policymakers have to ultimately assess the threat posed by an anti-American leader with control over a big part of oil production and reserves. Venezuela's military is no menace whatsoever, not even regionally, and the United States remains a big consumer of Venezuelan oil.

---

[1]IEA press release in July 2011: "A number of uncertainties remain which demand vigilance, notably the duration of the Libyan disruption, the future evolution of OPEC supply as well as the final impact of the stock release itself; much of the oil is only now entering the physical market."

Anything that further delays Venezuelan production this decade will have a big impact on oil prices and indirectly hurt America's economy. The United States would not be well served by further escalating a confrontation with Venezuela unless it concludes Chávez poses a security threat equal to that of Iran, an unlikely scenario any time soon. But in reality, American domestic policy is the driver of relations with Latin American, and over the past months there have been growing calls to confront Chávez on high military spending, geopolitical meddling, and his ties to China, Iran, and Russia, as well as terrorist groups.

Nobody expects Chávez to be friendly, or even conciliatory. A big part of his grassroot political support is built upon anti-Americanism. It would simply be political suicide to seek closer relations with the United States. And as long as Chávez limits himself to demagoguery and doesn't threaten American security, he can be managed. Besides, American bashing is a sport around the world, from China and Russia, to friendly countries in Europe and Asia. Furthermore, Chávez has always known that if he threatens American security, the playing field changes drastically and immediately. And so far he has always walked a fine line, making sure not to cross it.

Accurate data about Venezuela's weapons procurement is hard to confirm, especially because Chávez makes a public display of all his deals, many of which never materialize. But Venezuela's military spending was at least $43 billion between 2000 and 2010, only slightly higher than during the 1990s. Most of the expenditure of $32 billion was between 2004 and 2010, when the country replaced its obsolete arsenal following the American-imposed arms embargo in 2006, according to the Stockholm International Peace Research Institute, a bellwether tracking international arms expenditures.[2] The new hardware includes mostly Russian-made air defenses, radars, tanks, light weapons, fighter jets, war ships, and more. Washington is concerned, not so much because of its military hardware, but because discarded older weapons could be diverted to terrorists.

Venezuela still spent far less proportionally and in absolute terms over the same period than most peers in Latin America, including Colombia, Chile, and Brazil. Colombia spent $75 billion, Chile $54 billion, and Brazil $238 billion. As a share of annual GDP, Venezuela spent an average 1.4 percent between 2000 and 2009, one of the smallest ratios in South

---

[2]SIPRI data base: www.sipri.org/databases/milex

America. It compares to 3.6 percent of Chile, 1.6 percent of Brazil, and 3.5 percent of Colombia.

Furthermore, the spending is far from translating into military might because it will take years for Venezuela's forces to master the new Russian and Chinese hardware and even to bring its standing army readiness on par with countries like Colombia, Brazil, and Chile. Not even Venezuela's neighbors are concerned about its capability, especially considering internal dissent.

A secondary concern has been Chávez's geopolitical maneuverings to gain regional clout. Aside from his Bolivarian block, governments throughout the region have monitored whether Venezuelan oil diplomacy destabilizes their countries. It has at least been disruptive, especially in the case of Honduras and Nicaragua, where Chávez's oil checkbook has allowed Bolivarian-type movements to prosper. It could be through free eye treatment or foreign aid that ends up in political campaigns, as is believed to be the case in Argentina and Colombia, or just enough cash to subsidize goods to favor his allies.

Then again, governments and interests groups have been legally meddling in Latin America for decades. Overt aid is a legitimate form of foreign policy. Covert aid is not. And even the United States is a beneficiary of Venezuelan oil handouts. Venezuela has also protested against America's funding of Chávez's opposition. Cash diplomacy was not invented by Chávez.

However, Chávez's sway is overhyped by American media. Being Chavista is not a compliment in South America and in fact scares away voters, even if a statist trend continues. Ecuador is a Bolivarian country, but President Correa is increasingly implementing a lighter version. Peru, too, with its new President Humala, shares some of the same nationalistic and populist policies, but it has openly distanced itself from Chavismo. Venezuela's influence is circumscribed to smaller Caribbean nations and is mostly harmless in the broader scope of things, although still irritating.

Venezuela could thus only pose an asymmetric threat, and that is America's main concern. The government of Chávez has sympathized and given different levels of support in the past to groups considered terrorist organizations by the Department of State, especially the Revolutionary Armed Forces of Colombia, or FARC for their Spanish acronym, and ETA, the Spanish Basque separatist group. A second source of concern—and one directly related to terrorist ties—is the strong relations forged between Chávez and Iran, which have given to speculation about covert ties to Hezbollah of Lebanon and Hamas of Palestine, both proxies of Iran. The question is whether his ideological and material support constitutes a terrorism threat, a very high bar to meet by international standards, especially considering the consequences.

Chávez has definitely meddled in Colombia's civil war, playing both sides to his advantage. There is a trove of information documenting different levels of logistical support for the FARC during his administration, including multiple testimonies from former guerrillas and even former military officers in Venezuela. Most of the meddling though was to accommodate his own domestic political agenda and to strategically position Venezuela as a go-between in Colombia's war.

The FARC took advantage of peace initiatives that started in 1997 and ended in 2002 to consolidate their military, political, and geopolitical gains. A weakened Colombian state resorted to irregular paramilitary warfare to contain the guerrillas and drug mafias simultaneously, and increasingly many worried regionally that the government was losing control. At least 10,000 Colombians were killed, hundreds of thousands were displaced or injured, as multiple armies battled for territorial control, mostly at the hands of government-allied paramilitary armies that purged vast swaths of guerrilla-controlled areas, including the country's biggest cities.

When Uribe rose to power in 2002, he robustly strengthened Colombia's armed forces and put them on the offensive. He signed a controversial peace deal with paramilitary forces that has been criticized internationally as a blanket amnesty for gross human rights violations that accomplished little. Colombia's courts are still investigating paramilitary ties within the Uribe government. Indeed, the country's powerful intelligence service DAS was completely gutted in 2011 after authorities concluded paramilitary forces had all but taken over the agency (Peñate was not involved in DAS by that time). Several officials who ran DAS under Uribe have international arrest warrants for paramilitary ties, and some of Uribe's top aids, political allies, and family—but not Uribe himself—are being investigated or have been prosecuted for the same. Even the American Embassy was targeted by the DAS, as were most of the country's leading journalists, politicians, human rights advocates, and union leaders.

That said, most paramilitary leaders ended up killed or extradited to the United States on separate drug charges. Political violence drastically decreased with Uribe and the economy grew at unprecedented rates. Paramilitary armies lost their strategic gains once they splintered and the FARC and other guerrillas retreated back into the jungles, and both militant sides are struggling to regain influence and legitimacy.

In the middle of all this, Uribe asked Chávez to broker talks with the FARC. But it increasingly became clear that Venezuela was using the conflict to bog down Colombia and to further its own domestic and geopolitical interests. It was playing both the FARC and the government to its advantage.

Chávez's ties to the FARC were exposed with Colombia's capture of a treasure chest of documents seized from one of the FARC's top leaders during a cross-border air assault in Ecuador, March 2008. The leader went by the alias Raúl Reyes and was the second-in-command of the FARC. Reyes was the FARC equivalent to a foreign minister. And when Colombian planes bombed his jungle hideout across the border in Ecudor with the help of US intelligence, nobody knew they would recover decades of inside records of one of the world's longest-surviving rebel groups.

For Colombia, the incursion in Ecuador set off a regional diplomatic backlash from which it recovered only when Santos took office, but from a military and tactical point of view, it was one of Colombia's biggest battle victories. The FARC suffered tactical and operational blows they have yet to recover from.

The archive, which included eight mass storage devises in metal cases, was analyzed by the London-based International Institute for Strategic Studies (IISS), which presented its findings in a 240-page dossier in May 2011. The IISS concluded that the FARC and Chávez did indeed have active relations, as have all recent Venezuelan governments.

Chávez's main goal was geopolitical, rather than military. Chávez appears to have flirted with asymmetric warfare, but the FARC ultimately turned out to be unreliable and a handicap for his plans. On several occasions the guerrillas also betrayed Chávez, killed Venezuelan military personnel, and kidnapped Venezuelans for ransom. His ideological sympathies led to two short-lived rapprochement efforts that ended badly for Chávez, at times because rogue elements of his fractured government took things into their own hands, including top officials in the ministerial level and military hierarchy.

As far as the IISS could conclude, Chávez probably paid $240,000 for the release of Venezuelan and international hostages, something he denied and instead portrayed as a diplomatic success. But despite numerous promises of other financial support, military aid for the FARC didn't materialize. Chávez appears to have been goading the FARC with a decade's worth of failed promises and the FARC leadership didn't trust him much, even if it was hopeful he would eventually come to their rescue.

Venezuela did offer refuge and even a conduit to further FARC diplomacy abroad, apparently in an effort to keep the FARC close, but not too cozy, and to build an asymmetric deterrent to perceived US hostility. But Chávez was not reliable either, and when it was to his advantage, he had no problem sacrificing the FARC. In fact, the undercover capture in December 2004 of one of the top FARC leaders in Venezuela appears to have been covertly aided by Chávez in retaliation for attack on his border patrols. Logistical support appears to have been tolerated by rogue elements of his government. And Fidel Castro reportedly warned Chávez not to grow too close to the FARC.

Below are some excerpts from the IISS dossier:

> The FARC's involvement in Venezuela, sometimes assumed to have begun when Hugo Chávez took power in 1999, actually started much earlier.
>
> It is clear that Chávez has sought a relationship with FARC because of a convergence, in the most general sense, of strategic interests. Since the mid-1990s Chávez's and FARC's respective 'Bolivarian' discourses have had much in common, particularly their conception of Latin America as a collective of affiliated nations requiring transformation both in terms of their domestic politics and economics and their place in the international system.
>
> The Venezuelan president has worked to balance US influence in Latin America by seeking to build a series of alternative trade and political alliances, particularly through ALBA and by helping through various means to bring to power a new generations of friendly leaders. This allegiance-building has also had a marked extra-regional dimension, as Chávez has sought to build military and economic ties with Iran, Russia, and China. At the hemispheric level, Washington's heavy commitment to Colombian counter-narcotics and counter-insurgency efforts has made the Colombian conflict an appealing option for indirect confrontation with the US.
>
> And yet the commitment to FARC has proved fragile for two reasons. First, despite the overall strategic convergence, there is no

> firm ideological bond between FARC and Chávez, whose idiosyncratic and pragmatic approach to political problems has been perceived as incoherent and even alien by FARC. Secondly, the balance of power between Chávez and FARC has always been markedly unequal, with Chávez the stronger party. As a result, the president has not hesitated over the years to go back on promises made to the group, to distance himself from its, or even to harm its interests in Venezuela in pursuit of economic or political expediency.
>
> The FARC was able to operate in Venezuela and make use of Venezuelan resources without the permission or involvement of the Venezuelan government [through 2002]. It was inevitable that Venezuela's capacity to protect and control the long land border with Colombia would fall short. Chávez was well aware that violent conspiracy would be considered an option by dissident factions within the armed forces, and sought to avoid provoking these sectors unnecessarily.
>
> The Venezuelan government sought to control the scope of FARC's political and operational activities within the country [through 2002]. In particular, FARC's political activity required oversight and restriction, and this, combined with Venezuela's denial of major financial support, meant that FARC was frequently frustrated by the shortcomings of the relationship.
>
> Although it had failed, the [2002] coup not only dramatically heightened Chávez's perception of risk but also weakened his hold on power, meaning that his relationship with FARC, already poorly concealed, now constituted a potentially catastrophic problem for his domestic support and international legitimacy. Internationally, a relationship with FARC seemed increasingly risky, particularly with respect to US-Venezuela relations.[3]

Chávez was looking to heighten his international profile, but Colombia and the United States saw a covert effort to support the FARC. Both versions are true. Chávez hedged his bets by approaching the FARC. If it turned out well, he would be an international peace broker. If it turned out bad, he could always use those ties with the FARC as an asymmetric weapon, if needed. Chávez was becoming increasingly concerned by US-Colombian

[3]IISS FARC dossier, pp. 39–158.

intentions, especially with growing reports that Colombian paramilitaries were taking their war to Venezuela. Chávez saw a conspiracy, one that involved the United States and Colombia legally strengthening military ties, while covertly supporting irregular warfare.

Uribe and Bush made no effort to convince Chávez to the contrary. In fact, they used Venezuela's ties to the FARC—technically, nurtured at Colombia's request during peace talks—to expose Chávez, not always successfully. Eventually, Uribe denounced Chávez for meddling, with American support, which in turn opened the door to the signing of a broad military agreement with the United States. Chávez said that the two countries, especially Colombia, were a military threat. Colombia, which has received over $7 billion in military support since 2000, cleared the way for the United States to use Colombia as its main military base in South America, having been ejected from Ecuador. That is when diplomatic relations between the Venezuela and Colombia were frozen and not revived until Santos.

Andrés Peñate, the former Colombian intelligence head and strategic advisor of BP, analyzed bilateral relations.

> Chávez thought that Colombia under Uribe was just a US puppet, but it was never that way. That said, I don't think he was building an asymmetric threat, but rather playing to domestic opinion. Chávez believes that US strategy is to get rid of him, even from this world. I think there are influential people in his group who buttress this paranoia, like the Cuban intelligence services because their best interest is served by the US having a problem with Venezuela.
>
> Chávez is also interested in keeping Colombia bogged down. As long as the FARC are a problem for us, he can treat us as his younger brother.
>
> But I don't think that Venezuela's relation with the FARC was a type of Hezbollah relation with Iran. My reading is that the FARC allowed Chávez to pose as a left-wing radical revolutionary for Venezuela's traditional left.
>
> Besides, what's important in international politics is to create problems, not to solve them. You become an important actor depending on how much trouble you make. If Chávez were a US ally, nobody would notice him. It's a David vs Goliath agenda.

Regardless, what Chávez projected changed his relevance in the region to his benefit.

Uribe pushed through the agreement with the United States, which for Chávez confirmed that the US were out to get him. In 2009, Chávez mobilized troops and tanks to the border, which by any standard is a threatening military move, regardless of whether it was just a bluff that no one took seriously. Chávez might have calculated that he was sending a strong message that he would not accept Colombian support for a perceived threat from the United States. And it might have worked, considering Colombia backtracked on its military deal with the United States, as will be analyzed later. But the United States and Colombia will not soon forget Chávez's military response.

Still, despite relations to Colombian guerrillas, it has proven almost impossible so far to build a foolproof case against Venezuela as a terrorist sponsor.[4] Decades of war in Colombia fueled by a parallel drug war that involves the United States has had a spillover effect that historically and unavoidably involved all bordering nations and parts of the Caribbean. Basically, Colombia would need to prove Venezuela is a direct actor, which it's not, because Chávez can always claim, as he has, that he is just dealing with Colombia's mess.

Mark Weisbrot is codirector of the Washington-based Center for Economic and Policy Research (CEPR). He is economist and a regular columnist in global media, more than often critical of American economic and foreign policy, and has testified before Congress on Venezuela.

> A lot of what you see in Venezuela is inertia. Venezuela is a very weak State. Chávez doesn't have that much power. So to say there are people in the government who are friendly to [terrorism] doesn't really mean anything. It's an anarchist state, people do whatever they want.

A similar case can be made concerning Chávez's support for Basque terrorist group ETA, which killed 829 people in its 51-year struggle for independence from Spain before it declared a permanent and definite ceasefire in 2011. A Spanish extradition request for a militant who now works for the Venezuelan government has spiraled into a litmus test of Chávez's cooperation in the fight against global terrorism. At the heart of the simmering diplomatic standoff between the two historically friendly

---

[4]Country Reports on Terrorism, Venezuela.

governments is Arturo Cubillas, who was indicted by a Spanish court in 2010 for allegedly organizing terror-training workshops between ETA and the FARC in Venezuela.

Caracas ignored the international arrest warrant issued in 2010 for Cubillas and denied the extradition request issued by a Spanish judge. The indictment says the evidence "demonstrates Venezuelan governmental cooperation in the illicit collaboration" of the FARC and ETA, which included explosive-handling, intelligence sharing, and surface-to-air missile training seminars. If the case in Spain prospers, Venezuela's government could be legally incriminated, although that is unlikely now that ETA is seeking peace talks with the government. Venezuelan laws bar the extradition of its nationals. Cubillas, like several other former ETA militants, is now a Venezuelan citizen and thus shielded. Spanish authorities believe Cubillas is ETA's top militant in Latin America. He took refuge there in 1989 at the request of the Spanish government after failed peace talks. He is also the top security official in the country's Agriculture Ministry and is married to a high-ranking Venezuelan official in Chávez's government. But it would be even harder than in the FARC case to prove material terrorist support and the burden of proof is on Spain.

Ultimately, the biggest source of concern for the United States, and also the most unlikely, is the "brotherly" ties forged with Iran. Is Chávez allowing Hezbollah to train in Venezuela, or giving operatives passports to eventually reach the United States? Could he be giving Iran covert military support by diverting part of its newly bought arsenal, especially Russian air defenses? Or could he be exporting uranium to make bombs? And did Venezuela, Iran, and Cuba conspire to carry out a cyber attack on United States interests? Otherwise, what could a devout Persian Shiite Muslim possibly share with a liberal-minded Caribbean Christian?

Brotherly lessons from Iran

As we've analyzed, Chávez feels increasingly threatened by the United States and knows he could not confront it militarily. And this is where Iran comes in. The Islamic Republic has developed perhaps the world's most formidable expertise on asymmetrical warfare and has been in a constant state of war readiness, especially since American military superiority surrounded the country along most of its borders, including

Afghanistan, Iraq, and the Persian Gulf. The question is what exactly Venezuela has learned.

Iran and Venezuela have been bonding for well over a decade. Chávez has visited the Islamic Republic nine times, while his counterparts have traveled to Caracas eight times, most recently in January 2012. Soon after rising to power, Chávez found in Iran an ally to raise the price of oil, long before its nuclear program had been exposed. Iran at the time was gradually opening to the world and reforming its economy, but US sanctions were still biting and the country needed extra income.

Chávez was also eager to show off his revolutionary credentials. In 2000, he was the first head of state to visit then Iraqi President Saddam Hussein since the 1991 invasion of Kuwait. He crossed the Iranian border by land to reach the Iraqi dictator after getting a warm embrace in Tehran.

Iranian President Mohammad Khatami returned the visit in 2000 ahead of an OPEC summit in Caracas, and twice more before Ahmadinejad was elected in 2005, signing a myriad of agreements with Chávez's government. The Venezuelan leader also returned three more times to visit Khatami and Supreme Leader Ayatollah Ali Khamenei, the last in 2004 when he received the highest Iranian honors.

That was when Chávez met Mahmoud Ahmadinejad, a young Tehran mayor, an upcoming star in Iranian politics. The two often describe relations as "brotherly." All throughout, Chávez staunchly defended Iran's right to develop a peaceful nuclear program. Businesswise, bilateral trade is insignificant and a lot smaller than with other Latin American countries, compared to the multibillion dollar trade with the United States. In 2010, Iran's trade with Latin America was around $1 billion, 90 percent of which was with Brazil and Argentina,[5] and in any case only a fraction of its global business. Global economic turmoil and sanctions are probably to blame for unmet promises of at least $2 billion in committed Iranian investments in Venezuela in finance, industry, and energy sectors. The financial sector is perhaps the most sensitive as banking alliances could be used by Iran to circumvent sanctions and access credits. The Treasury Department imposed sanctions on an Iranian owned, Caracas-based bank in 2008.

Iranian presence in Latin America has been mostly limited to OPEC relations. Unofficially though, its proxy Hezbollah has been present

[5]IMF, through EU Commission global trade statistics.

throughout the continent for some time, especially in the border of Argentina, Paraguay, and Brazil, mostly for fundraising. Hezbollah was blamed for the bombing of two Jewish centers in 1994 that killed 85 people, although the group and Iran deny the charge.

But Chávez opened Venezuela's doors to Iran at least temporarily. It is impossible to know what kind of intelligence and military exchanges have taken place between both countries, especially since Ahmadinejad came to power in 2005. Chávez has visited his friend in 2006, twice in 2007, once in 2009, and again in 2010. Until 2010 there was a weekly flight connecting Tehran, Damascus, and Caracas. The US Department of Treasury froze the assets of two Venezuelans, including one diplomat, "for providing financial and other material support to Hezbollah."[6] The Pentagon has also advised Congress that the biggest concern with Iranian-Venezuelan relations were ties to Hezbollah.

Since 2006, the United States includes Venezuela in a list of countries "not cooperating fully with the US in antiterrorist efforts." But it's unlikely though that Chávez would risk blatantly supporting terrorism, simply because it could open his government to sanctions and confirm hawkish American fear-mongering that Chávez is a terrorist threat. So far there is "no corroborating evidence" that Hezbollah or Hamas have operative terrorist cells,[7] which any US administration would have long confronted if that were the case.

Indeed, the Iranian-Venezuelan axis appears to have diluted with the global economic crisis. Their common interests, in fact, are increasingly circumscribed to coordinating bullish pressure on oil prices, as we will see in Chapter 6. Perhaps Chávez was just testing the ground, or learning from his fellow populist, oil-wealthy brother in arms. Or perhaps American pressure was enough to make Chávez second-doubt any ill-intention he might have harbored. Or it could be that Venezuela and Iran are both cash strapped but will once again seek a more practical partnership in the future. More than likely though, the economics and politics were simply too big to overcome.

---

[6]www.fas.org/sgp/crs/row/RL32488.pdf

[7]Congressional Research Service, *Latin America: Terrorism Issues*, July 2009: "The State Department terrorism report maintains that the United States remains concerned that Hezbollah and Hamas sympathizers are raising funds among the sizable Middle Eastern communities in the region, but stated that there was no corroborated information that these or other Islamic extremist groups had an operational presence in the area."

The consensus in Washington, in the words of Negroponte, is the following:

> The rumor might be ahead of the story. My impression is that Hezbollah was in Latin America to raise funds, not to foment terrorist activity. It's not a threat to us. Can Venezuela be a thorn on the side of some of its neighbors? I think so. It has some hegemonic aspirations with the Bolivarian initiative and all of that, but I don't know who's going to be fooled by that.

A new Chávez?

In 2011 under a new Colombian administration, Chávez offered a much more conciliatory foreign policy, distancing himself slightly from the subversive persona, as we will analyze in depth in Chapter 9. He cooperated with Colombia against the FARC, meaning that he had at least temporarily renounced any asymmetric ambitions. Chávez probably concluded that if caught with a smoking gun, he would be shunned by all, even by Russia and China, not to mention a vast majority of Venezuelans, including his supporters, who don't want to be ruled by a global pariah.

While it's too early to call it a trend, Chávez in 2012 is also trying to attract foreign investment, as illustrated by the natural gas deal signed with European companies, which we'll analyze further in Chapter 7. PDVSA offered a significant market premium to the companies. It's encouraging that Venezuela can be pragmatic when it comes to foreign investment in its energy sector. It could signal a more business-friendly approach in the near future as Chávez increasingly struggles to make his revolution economically viable.

Regardless, Venezuela's economy is underperforming its peers, despite its huge oil production and sustained high oil prices. Like in many oil-producing countries, oil wealth is squandered and mismanaged. Chávez's most pressing priority ahead of elections, like most other incumbents, is the economy. At this juncture, he has little option but to move closer to the center, or risk losing the election and his revolution.

Despite high international oil prices, Venezuela's economy plummeted 3.2 percent in 2009 and an additional 1.5 percent in 2010. Crime rates have persistently increased. Inflation is the highest in the region and oil production, the country's livelihood, is still below pre-Chávez highs. In

2011, the economy grew 2.8 percent on the back of higher oil prices, but still trailed all its peers. Oil production additions are not offsetting depletion rates in older wells, although investment increased last year and will pick up in 2012.

Venezuela inflation was a staggering 26 percent in 2011, by far the highest in the region, followed by Argentina with 11 percent, according to the IMF. In addition, Venezuela is facing at least 18 cases filed in the International Center for Settlement of Investment Disputes against expropriations.

That said, Chávez has been able to secure enough resources using his country's oil production to underwrite his antipoverty measures, from housing to healthcare. He still loosely describes the macroeconomic failings as part of transitioning into 21st Century Socialism.

Chávez doesn't have a credible track record, at least economically. But that could be said of many oil producers. He still has more than enough resources to sustain the regime intact. Unless oil prices tumble for several years, Venezuela still has the region's fifth biggest GDP and a robust service sector.

Weisbrot, of the CEPR, about Venezuela's economy:

> It's very much an oil based economy in the sense that 95% of its exports are oil and the majority of government revenues. But it's less than most people think in terms of GDP. The vast majority of economy is domestic goods and services, although it's still an oil state by any international comparison.
>
> Business press has been predicting disaster year after year for the last 15 years. The disaster hasn't come except for one and half year recession. Even that wasn't so much of disaster. And it was avoidable.
>
> There's waste, yeah. But from an economic point of view, waste is not necessarily bad. Even corruption is not deadly to a developing economy. South Korea had two presidents go to jail for corruption. Within the country there is a redistribution of income so it doesn't matter so much. It's still bad, but it doesn't hurt your growth and development.
>
> Venezuela's problems are simple. They had a recession that was unnecessary and I don't think it will happen again. Objectively, they're recovering now. They really shouldn't have any trouble. A country sitting on so much oil can always borrow and can always do whatever is necessary to stimulate the economy, if private spending falls.

> Chávez talks a lot, but he's never said a word about defaulting. They're never going to default. Obviously there is political risk.

Former President Uribe isn't so optimistic, and most share his assessment:

> Security and private initiative have deteriorated in Venezuela. If there is no private investment dynamic, social policies are unsustainable, regardless of how much natural resources they have.

Venezuela's economy is still expected to underperform most Latin American peers, although higher oil prices could change that. The country's economy will grow around 3.6 percent in 2012, according to the IMF, although expensive nationalizations and social programs are increasingly being paid with debt. In August 2011, Standard & Poor's downgraded Venezuela for the first time since 2006, citing "changing and arbitrary laws" and Chávez's health. Now Venezuela is on par with Bolivia, Paraguay, and Surinam. Chávez also repatriated some $11 billion in gold reserves it held throughout the world, a move that analysts say will just undermine credibility on its economy. He also transferred some $6 billion in cash reserves from the United States and Europe to Russia, China, and Brazil. Lenders demand around 12 percentage points more from Venezuela than from the United States, compared to 9 percentage points more from Ecuador and nearly 2 percentage points more from Mexico.

"The heavy spending levels in preparation for the 2012 presidential election likely will take the general government deficit to 2.5% of GDP in 2011 and about 4% in 2012. Consequently, we expect that net general government debt will rise to 17% of GDP in 2011 and 19% of GDP in 2012, despite double-digit inflation boosting the denominator,"[8] S&P said.

Venezuela's gross debt as a percent to GDP was around 44 percent in 2011, according to the IMF, even after the steep hike in lending over the previous two years. Compare that to America's debt levels of 100 percent and the European Union's 80 percent. Part of the concern though is that Chávez is also relying increasingly on bilateral loans underwritten with oil production, especially from China, but also from countries like Japan and South Korea. In the meantime, Venezuela's oil output is at best flat, which implies its ability to pay is deteriorating, not improving.

---

[8]Standard and Poors. "Venezuela Long-Term Ratings Lowered To 'B+' From 'BB-' On Political Risk And Revised Methodology; Outlook Stable" August 19, 2011.

Bottom line, Venezuela's economy is deteriorating and could be in significantly better shape, but it's far from being an economy in chaos or even one at risk. In fact, the biggest factor influencing Venezuela's economic outlook is not its macroeconomic stability, but its political stability. Concerns are not about the economy per se, but about uncertainty in its economic management.

Chávez's ability to harness power around himself heightens uncertainty. In December 2010, for example, the Venezuelan parliament approved Chávez request to manage the economy by decree through mid-2012, bypassing legislators.

Uribe on Chávez's move:

> What happened is serious. Venezuelans elected a Congress and then the government took away its power before it could take office. That to a certain extent is a coup on one of the branches of government. We talk about coup d'etats, but not of governmental coups. The government gave a coup on the legislative branch.

His decision though highlights reality. Chávez can get away with his constant undermining of Venezuela's democracy, despite intense criticism inside and outside. He is not really concerned by the opposition ahead of next year's elections or of any foreign threat, especially now that its relations with Colombia and the rest of Latin America have improved.

That explains why Chávez is confident enough to relax his hostile rhetoric and to spend long periods of time in Cuba, even after two years of recession. He has the money and power to govern without any real check. And he has pursued more regional integration by toning down his militancy, all during the public exposé of his struggle against cancer, a drama that is evolving in a parallel narrative to his government.

His apparent U-turn is very recent and ahead of presidential elections, which could suggest a tactical retrenchment. But an easier explanation is that he is adapting to a new reality in Venezuela and the region. If indeed Venezuelans wanted radical change a decade ago, they are now demanding more stability, and Chávez seems to be moving however temporarily to the center.

The question at this juncture is whether Chávez can fulfill his promise of doubling Venezuelan production. After all, it's America's national security that is at stake.

PART TWO

THE CENTURY OF OIL, AGAIN

4 AMERICA AND ENERGY: AN "ADDICTION" HARD TO KICK

Is the world running out of energy? The simple answer is no. There is a huge difference between depletion and increased competition. There is more than enough energy from known sources to meet demand for decades, if not centuries to come. And considering our rate of innovation, it's very unlikely, if not impossible, that it will ever come to a point when there is not enough energy to go around, barring of course a cataclysm.

Oil is not peaking any time soon, nor is any other fossil fuel, despite all the rhetoric. If nothing were to change, nothing at all—no new deposit discovered, no technology developed, and humans consumed like they do today—there would still be enough coal to last the world three centuries, enough oil for another century, and enough gas for several centuries.

But things will change, as they always have. More fossil fuels will be found; the market share of natural gas will continue rising; new technologies will be developed, and countries and citizens will alter their consumption trends, with vast differences from China to Paraguay, to the United States. Here's how the IEA described misconceptions about peak oil in its World Energy Outlook (WEO) 2010, which looks out to a 2035 horizon.

> The amount of oil that was ever in the ground—oil originally in place, to use the industry term—certainly is a fixed quantity, but we have only a fairly vague notion of just how big that number is. But, critically, how much of that volume will eventually prove to be recoverable is also uncertain, as it depends on technology, which will certainly improve, and price, which is likely to rise: the higher the price, the more oil can be recovered profitably. An increase of just 1% in the average recovery factor at existing fields would add more than 80 billion barrels to recoverable resources (IEA, 2008). So, the chances are that the volume

> of resources that prove to be recoverable will be bigger than the mean estimate we use to project production, especially since that estimate does not include all areas of the world. Even if conventional crude oil production does peak in the near future, resources of NGLs and unconventional oil are, in principle, large enough to keep total oil production rising for several decades.

The ways we relate to energy will also change, like they always have. For centuries, the world was hooked on coal for steam and heat as its primary energy source. Then came mass distribution of electricity and automobiles, which prompted a shift to petroleum in the transport and heating sector and, increasingly, nuclear and natural gas in power generation. All these revolutions are the result of the world's ever growing hunger for affordable and secure energy.

In coming decades, the world will surely accommodate once again as available resources are reassessed, and as new ones are developed. The nuclear age is not even a century old and is sure to expand, despite the Fukushima meltdown in Japan in March 2011. Renewable energy sources have only begun being tapped aggressively over the past decade and their contribution will continue to grow, and even if their share remains comparatively small, they are nonetheless a vital component of the electricity sectors in some markets such as Europe. Perhaps over the next few decades promising breakthroughs using nuclear fusion or hydrogen could well fuel energy revolutions that end the reign of fossil fuels. But science fiction aside, for two decades at least, the world's machines will run mostly on fossil fuels. America and the rest of the world will surely find ways to meet demand, for this and future decades. The questions are, at what cost, and under what conditions?

America's energy future can be explained as per capita energy demand, that is, how much primary energy each person consumes on average, which drives the public debate. *Energy intensity* though, that is, how much energy is consumed to produce one dollar of GDP, drives the decision-making process.

If energy prices increase too much and the United States cannot secure cheap enough resources to keep its economy competitive, more jobs will be lost, more industry will relocate, and the trade imbalance will only widen. But reducing energy demand on a per capita basis is hard and politically unattractive, and significantly harder than reducing the energy intensity.

Ultimately, the American roadmap to improved energy security, regardless of how it combines its different supply sources, is to increase local supply and cut down on demand while the population rises and the economy grows. In a perfect world, a transition would allow markets to balance out the economy. But since America's ability to boost energy supply domestically is naturally constrained, increasing output elsewhere is the only option. In this decade, that translates into vastly more global oil and gas production capacity, on top of more electricity generation, which will vary geographically as countries balance nuclear, coal, or renewable power with economic growth.

That is why America's blind spot is not a futuristic guessing game. The analysis is based on challenges the county is already facing. After 2020, energy debates will no doubt be very different as geopolitics and technology redefine trends. This is about realistic and feasible options that will not only secure energy supplies, but also drive energy prices down to pay for our way of life.

Global energy trends

The problem with projections is that they vary depending on the assumptions, themselves often colored by ideology, whether it's global economic growth, population, or government policies implemented over time. Historically, long-term forecasts have proven mostly unreliable, able to point to trends, but off the mark when it comes to real consumption or production. The unpredictable nature of geopolitical instability and technological advances is to blame.

Uncertainty is part of the game of course and it's not new. How could forecasters factor in a Greek exit from the European Union, a war with Iran, or an accident like the Macondo one in the Gulf of Mexico? What all can agree on is that, at this juncture, price hikes are more dangerous than ever. There is very little room to maneuver this decade when it comes to balancing supply and demand, and even less when it comes to the global economy's ability to sustain large swings in prices. The last five years have witnessed some of wildest fluctuations, and forecasters have been forced to recalibrate their expectations from one month to the next. Worse yet, things won't get better any time soon.

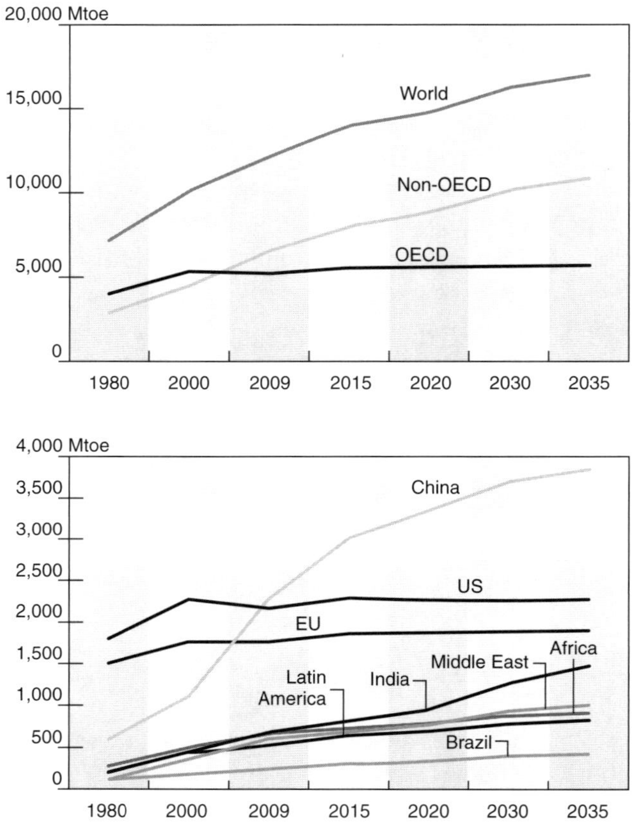

FIGURE 4.1 World primary energy demand by region or country
Source: IEA WEO 2011

The world's primary energy demand, which includes all forms of energy, increased by around 45 percent in the past two decades. In the next 20 years, demand growth is expected to slow down, despite an increase in population and robust economic growth as developing nations like China and Brazil catch up to industrialized countries' consumption trends.

Globally, the population will increase less than 1 percent annually through 2035. Europe's will increase only slightly, while Latin America's and North America's will see moderate growth. World economic growth will be vigorous, at a rate of more than 4 percent annually through 2020 and slowdown thereafter. The economies of non-OECD countries, driven by China and India, will expand by around 6 percent annually through 2020, while industrialized nations will do so at about a third of this rate.

Most forecasters assume stable energy prices, benchmarked to crude oil, fluctuating between $100 a barrel and $130 a barrel throughout this decade. A general trend, certainly in the United States and other OECD countries, is a gradual convergence in the demand of fossil fuels to account for availability. That translates invariably into less crude oil, a fluctuating, but stable coal demand, and more natural gas.

Oil, coal, and natural gas will continue supplying more than three-quarters of global energy for at least two decades, and certainly through 2020 when coal and oil alone will account for 60 percent of all energy.

Hydroelectric power and nuclear output will remain largely flat, while the renewable energy contribution will rise significantly—albeit at a huge expense to taxpayers—but the combination of all these sources will still struggle to supply around 20 percent of world energy use through the remainder of the decade, and only slightly more by 2030.

Fossil fuels will thus continue ruling the energy scene for several decades, despite growth from all other sources, which in any case will depend to a great extent on government subsidies. World energy prices will continue to be benchmarked to oil markets, and economic growth will continue to be largely conditioned by its fluctuating market.

In other words, global and US economic growth hinges more than ever on the balance between oil supply and demand. And considering that easy to extract oil—commonly known as conventional oil—is rapidly running out, production of the more expensive and harder-to-produce oil—or unconventional oil—will have to rise.

America's "addiction" to energy

Total energy demand through 2020 increases by nearly 7 percent in the United States, and 13 percent by 2030.[1] Fossil fuels will continue supplying around 80 percent of total energy over the next two decades, despite a 10 percent increase in nuclear output and 38 percent increase in renewable energy, which includes hydroelectric power.

---

[1]EIA's Energy Outlook 2011 report.

But the United States stands out from most other countries, developing or industrialized, in its voracious energy appetite. In per capita demand terms, every American in 2008 roughly used 400 percent more energy than the global average, the highest in the world by far. It is trailed by Russia, a notoriously inefficient and subsidized economy with a per capita demand that is about 250 percent higher than the world's average. Japan, America's closest peer, uses around 200 percent more than world average on a per capita basis, and the European Union around 190 percent more. That is, every American consumes about twice as much as their European and Japanese peers, as Figure 4.2 shows.[2]

The United States has more vehicles on the road on a per capita basis than any country in the world, including fellow OECD countries. In terms of overall electricity consumption, it is second only to China, which has four times more people.

Many people argue that Americans would be willing to significantly cut their energy demand. After all, many say, millions of Chinese lack

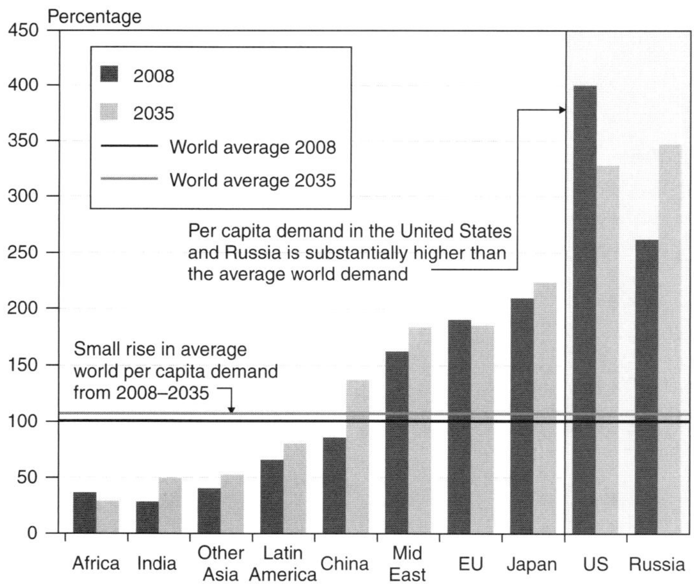

FIGURE 4.2 Per capita primary energy demand by region
Source: IEA WEO 2010

[2]IEA 2010 WEO.

power all together, so it should be easier for Americans to simply drive less and carpool more, for example. But the political reality is different, in the United States and China alike. It is America that China holds as its role model, and they want to elevate their lifestyle to that of America's. This will amount to one of the biggest challenges for the world over the next several decades.

Take internet access, for example. Today it's indispensable for Americans. It's in cell phones, video games, cafes, restaurants, airports. It's everywhere and everything runs on it. But two decades ago, hardly anybody had an email account. And that is limited to a 20-year time span. To ask Americans to change a lifestyle that has been sustained for two generations is nothing short of revolutionary.

Things are changing, albeit very slowly. "U.S. energy use per capita was fairly constant over the 1990 to 2007 period, but it began to fall after 2007. Energy use per capita continues to decline due to the impact of an extended economic recovery and improving energy efficiency," according to the preliminary estimates released in January 2012 by the EIA.

Into the future, the United States will also improve its energy intensity, and by 2035 Russia is expected to be the biggest per capita consumer, followed by the United States, Japan, and Europe. Surging developing economies, especially China, will increase the average global per capita consumption. "Energy use per 2005 dollar of GDP declines by 42 percent from 2010 to 2035 as the result of a continued shift from manufacturing to services (and, even within manufacturing, to less energy-intensive industries), rising energy prices, and the adoption of policies that promote energy efficiency," according to the preliminary EIA 2012 forecasts.

America's voracious energy demand, and its political inability to clamp down over the decades, is not problematic per se. Indeed, human progress and energy demand are correlated. The problem is that over the decades the US economy has grown dependent on plentiful, cheap, and secure energy resources. The world's relation to energy has changed drastically though. American energy is not as plentiful, as cheap, and certainly not as secure. That is why energy independence is back in media headlines and politicians' speeches. But it's unrealistic to change consumption trends, at least rapidly, and it's not even advisable because, at the very least, it amounts to asking Americans to alter their lifestyles considerably. Policymakers fear the prospect of supporting the reduction of energy consumption more than raising taxes, and justifiably so. Some have tried and paid (or will pay) the ultimate price.

One reason, perhaps the most important, that Americans consume so much more than their European and Asian peers is that while those postwar, cash-strapped nations were struggling to secure enough energy to feed their recovering economies, the United States married itself to cheap indigenous production, tapping into its abundant resources, especially oil, coal, and gas, and later nuclear. But especially oil. Another important reason is of course geography. It is hard to compare the energy consumption of the United States and, for example, the Netherlands, a country that one can drive across in two hours.

As crude demand increased, the American response was to boost local production, which it managed to do for decades at affordable prices. Before 1950s, the United States was a net exporter of oil. Up to 1973, before the Arab embargo, oil imports accounted for slightly more than 30 percent of demand, but they increased around 65 percent annually through the rest of the 1970s. However, the embargo, and the Iranian revolution and embassy hostage crisis exposed just how vulnerable the United States was to oil flows.

President Jimmy Carter warned that a day would come when oil would be harder to obtain and more expensive. His energy reform policies, which included deregulating oil and gas prices, the creation of the DOE, strict conservation and efficiency guidelines, and the introduction of oil import quotas, propped up local production and innovation to gain oil independence. He also increased windfall taxes on oil companies. US oil imports shrunk for the first time in history thanks to Carter's policy, creating a global oil glut that eventually sent prices plummeting.

Ronald Reagan reversed many of Carter's policies and liberalized energy markets. But he cut taxes and increased company revenues, increasing oil production for some time and slowing down a negative trend. That in turn increased exploration and oil recovery techniques. The price of oil plummeted to less than 10 dollars a barrel, even while absorbing a major geopolitical supply disruption like the Iraq-Iran War.

Energy policy was forever changed by both Carter and Reagan. Looking back, their policies shaped American foreign policy and energy security. Carter's top priority was to make it clear globally that America could be weakened, but not crippled by its dependence to oil.

Reagan profited from Carter's energy policies, but his priority wasn't energy. It was the Soviet Union. Of course, Reagan could have extended Carter's policies that drove down oil imports and increased energy independence, but he had a bigger fish to fry. Reagan understood that the only way to defeat the Soviet Union was to drive it into bankruptcy. He did that by spending billions on nuclear weapons and military hardware, forcing the Soviets to do the same. The strategy could only work if oil prices remained at record lows though. The Soviet Union's bloodline was its oil exports. It remains so for Russia today. When prices plummeted, so did its ability to sustain its communist model. The opposite was true for the United States. The lower oil prices were, the more its economy would benefit in the short term. While the United States was in the middle of a bonanza, the Soviet Union was crumbling economically from within.

Policies don't exist within a vacuum. A combination of Carter's energy policies and Reagan's foreign ones had the unintended effect that energy security fell off the radar screen as Americans entered two decades of cheap oil. Reagan would not have been able to defeat the Soviet Union without the legacy of Carter's energy policies. But the consequences defined the administrations of George Bush, Bill Clinton, George W. Bush, and of course Barack Obama.

By 1994, America was importing more than it was producing, a trend that only recently started being reversed with new techniques. Oil imports increased for two decades, more than doubling since 1985, only to start falling again in 2005, after prices hit record highs. Increased domestic contribution though is naturally limited. In fact, the EIA now expects US crude imports to decrease significantly into the future, but mostly as a result of lower demand—as opposed to higher local production—in the aftermath of the global economic crisis.

Geopolitics and energy

Geopolitics is increasingly a major factor in energy prices and availability. The United States institutionalized energy security as a vital interest in 1980 when it committed itself to protecting its national interests in the Persian Gulf, when the issue was more about physical supplies than price. Here's President Jimmy Carter defining America and its energy security:

> The region which is now threatened by Soviet troops in Afghanistan is of great strategic importance: It contains more than two-thirds of the world's exportable oil. The Soviet effort to dominate Afghanistan has brought Soviet military forces to within 300 miles of the Indian Ocean and close to the Straits of Hormuz, a waterway through which most of the world's oil must flow. The Soviet Union is now attempting to

> consolidate a strategic position, therefore, that poses a grave threat to the free movement of Middle East oil.
>
> This situation demands careful thought, steady nerves, and resolute action, not only for this year but for many years to come. It demands collective efforts to meet this new threat to security in the Persian Gulf and in Southwest Asia. It demands the participation of all those who rely on oil from the Middle East and who are concerned with global peace and stability. And it demands consultation and close cooperation with countries in the area which might be threatened.
>
> Meeting this challenge will take national will, diplomatic and political wisdom, economic sacrifice, and, of course, military capability. We must call on the best that is in us to preserve the security of this crucial region.
>
> Let our position be absolutely clear: An attempt by any outside force to gain control of the Persian Gulf region will be regarded as an assault on the vital interests of the United States of America, and such an assault will be repelled by any means necessary, including military force.[3]

The policy known as the Carter Doctrine remains in place, even if the Soviet Union is no longer a threat. Reagan extended the doctrine to include threats inside the Persian Gulf, thinking of Iran. But Saddam Hussein suffered the ultimate consequence for his invasion of Kuwait. He threatened free movement of oil from the region and the first Gulf War was a direct application of the Carter Doctrine.

Today the biggest threat is Iran. If attacked it could try to shut the Strait of Hormuz, the most important oil passage in the world where 40 percent of the seaborne oil exports transit through. American military might is already in place to prevent that from happening. The military buildup would appear disproportional considering America only imports around 1.5 million bpd of oil from the Persian Gulf. That is around 15 percent of all oil imports, less than 10 percent of its oil and other liquids supply, and only 3 percent of total energy demand.

The reason that the United States protects the Strait of Hormuz at all cost is the price of oil. A prerequisite to global economic growth is for oil to flow freely and for prices to reflect market-based supply and demand considerations. Wars have been fought and will be fought to make sure

[3]Jimmy Carter during the State of the Union address, January 23, 1980.

energy flows. Washington told the Soviet Union three decades ago, Iran today, and will tell any country in the future that energy is critical to America, and securing global oil flows is as important as containing nuclear proliferation, perhaps even more.

The United States wouldn't collapse in the event of a temporary interruption to Persian Gulf oil supplies. But the world couldn't live with an oil shock for an extended period. The strategic oil reserves held by the United States would run out in a few months if it had to rely on local production and reliable imports from Canada and Mexico.

That scenario, one where oil supplies would be disrupted for so long, is farfetched though. Still, the underlying concern, especially when countries like Iran threaten oil supplies, is price. By putting the US military as guarantor of global oil supply routes, America allows energy markets to concentrate mostly on supply and demand.

The new concern for markets is increasingly the limited global spare capacity, that is, the amount of oil measured in barrels per day that the world is physically able to produce at a given time minus what is actually consumed. A big component of why oil prices soared throughout the past decade was the unexpected reality following the American invasion that Iraqi production would not be a game changer any time soon. Then Iran's nuclear program was exposed and oil markets guessed whether a face-off with the United States was possible. And, as that issue was put on the backburner amid the economic crisis, the Arab Spring took hold. The most pressing question became whether big oil producers in the Persian Gulf would ever see their exports affected. But even before the supply risks could be addressed, war mongering in Iran and Syria picked up again, along with instability in Algeria and Sudan, pushing oil prices higher.

Energy prices, the real threat

Energy security, especially for the United States, has become more a question of price, a trend that will deepen throughout this decade. Historically, surging oil price with few exceptions have led to American recessions, such as those that followed the Arab oil embargo in 1973, the Iranian hostage crisis in 1979, the invasion of Iraq in 1991 (although a credit crunch back then also weighed heavily), the OPEC production cuts of 1999, and of course the most recent recession, when the financial crisis was compounded by record prices of almost $150 a barrel.

It's not as easy as equating high oil prices to recession. Each downturn is its own monster. But there is broad consensus that economic growth is directly related to oil prices. Indeed, inflation tracks oil prices. And economic growth is inversely proportional to inflation, broadly speaking. That is, the higher energy prices are, the less money Americans have to spend on everything else, which means the economy slows. "Not only do oil and other energy prices constitute a portion of the actual Consumer Price Index, but downstream impacts on other commodity prices will have a lagged effect on the CPI inflation," the EIA said in a study of the relation between economic growth and oil prices (see Figure 4.3).

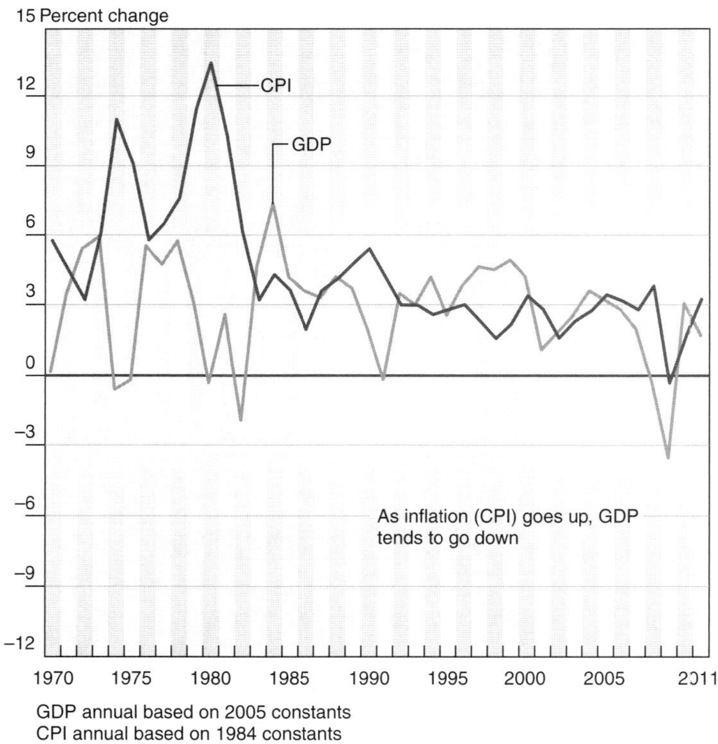

FIGURE 4.3 Oil prices and US economic growth

Source: US Bureau of Labor Statistics for CPI and Bureau of Economic Analysis for GDP

Every $10 increase in the price of oil reduces US GDP by 0.2 percent, according to a report published by the Congressional Research Service in April 2011, entitled U.S. Oil Imports: Context and Considerations.[4]

> Increased fuel costs for households and businesses can reduce spending on other goods and services that might be domestically produced, send more wealth abroad, and cause economic dislocations, such as greater unemployment. Economic forecasters estimate a sustained $10 increase in the per barrel price of oil can reduce U.S. economic growth by 0.2% in part due to how much U.S. consumption is met by imports. However, even in a scenario where the U.S. produced as much oil as it consumed, an increase in international oil prices would still raise oil costs for households and businesses and cause economic dislocations, but wealth associated with the increase may remain in the country.

Record high oil prices in the past few years, and renewed bullish trends in 2012, remind us that energy security is vital to the United States. It's no longer so much about oil embargoes or the ability to find resources, it's about prices.

How long can the US economy sustain $150 a barrel oil prices, or $200 a barrel, when half of the country's oil supply is bought overseas? And how many leaders are willing to accept the political cost of rising oil prices? For every $10 increase in the price of the barrel of oil, a gallon of gasoline in the United States increases 24 cents, inflation increases by almost 0.4 percent, real disposable income decreases by around 0.25 percent, and Americans spend about $30 billion more, according to IHS, the global energy consultancy. If those prices are sustained over one year, 120,000 jobs are shed and the GDP shrinks by about 0.2 percent. If the prices are sustained over two years, 410,000 jobs are shed and GDP shrinks by about 0.5 percent.[5]

IHS expects a "noticeable jolt to consumer sentiment" with S3 and $4 a gallon pump prices, a rapidly approaching threshold.

> In the present context, the most obvious candidate for a "tipping point" in the United States would be $4/gallon gasoline prices. It

---

[4]www.fas.org/sgp/crs/misc/R41765.pdf

[5]IHS report from February 2011: www.ihs.com/products/global-insight/industry-economic-report.aspx?id=1065929077

> would probably require another $20/barrel added to the oil price (i.e., around $120/barrel for WTI or around $130/barrel for Brent) to take gasoline prices through the $4/gallon mark this summer driving season. Compared with where we began the year (around $95 for Brent) that would mean a drag on GDP growth of 0.7% from the $35 hike (three and a half times 0.2%), plus whatever extra damage might kick in from the "threshold effect" of passing the $4/gallon mark. At that point, the damage to growth would start shifting from an irritant to something worse.

The global economy can sustain oil prices that represent about 9 percent of GDP, a level it only reached right before the double-dip recession in the early 1980s and then again shortly before the Great Recession of 2008, according to a Bank of America–Merrill Lynch report published in February 2012. In 2012 that would translate into $130 a barrel of Brent crude, which has hovered above $100 in 2012. It would take an unexpected slowdown in global growth to dent those forecasts, according to the report.[6]

The American economy thus depends heavily on keeping oil as cheap as possible and since demand is harder to tackle, supply is the only real way to influence prices. Fear of oil price volatility also explains the contradictory American reaction to popular uprisings in the Persian Gulf. Rather than supporting democratic reform as it did in Egypt, Tunisia, Libya, and Syria, the United States made it clear that its top priority was stability, even if it meant undermining pro-reform groups. The United States could simply not deal with revolutions spreading to the Persian Gulf, not because it likes supporting antidemocratic governments there, but because a regular flow of oil trumps all considerations.

Imagine if there was a disruption to Iranian exports? Or instability in Russia? Or even Saudi Arabia? The current spare capacity would disappear if Iran couldn't export crude, for example. That alone would send prices through the roof and the world into chaos. Any long-term disruption would be impossible to meet. The Libyan war put OPEC to the test and it failed.

---

[6] *Oil Demand Scenarios for 2012*, Bank of America–Merrill Lynch Report, October 11, 2011.

Long-term pipedreams

American policymakers left and right and business leaders flaunt energy independence as if it were a lever that governments refuse to pull.

Energy independence is not something the US could achieve easily, cheaply, or soon. By 2020, the United States will import less than 40 percent of its total supply of oil and others liquids, a significant drop from 49 percent in 2010, thanks to increased domestic production of crude and other liquids, and to a flat fuel demand, according to the EIA's early release of its 2012 Annual Energy Outlook (AEO). There is little change though to the global oil supply and demand balance, which means it won't have a big effect on prices. Furthermore, the United States is in no economic shape to even consider paying for energy independence. Not even China, with the world's biggest currency reserves, is considering such a pipedream. The best to hope for, which is true of most countries, is for increased energy security.

"There are no short-term solutions to our long-range problems. There is simply no way to avoid sacrifice," Carter said in a televised speech in 1979.

"There are no quick fixes. We will keep on being a victim to shifts in the oil market until we finally get serious about a long-term policy for a secure, affordable energy future," Obama said in 2011.

Politics has handicapped US energy security for decades and continues to do so. Short-term policies have dominated Congress and all administrations since Carter and Reagan. The Obama administration has indeed taken shy steps. In March 2011, Obama set the goal of cutting oil imports by 30 percent over 10 years, which is already on track, according to the EIA's 2012 forecast. Obama's response addresses origin of supply issues, but the priority into the future is decreasing prices.

Will Obama make any hard choices during an election cycle? Hardly. So far, this administration, like so many before, has offered lukewarm, short-term energy policies, like promising to ease exploration restrictions, but not everywhere and not quickly enough. Public funds will go to developing alternative sources, but not nearly enough to alter the equation in the short term. Indeed, policies are in place to avoid worsening the problem too much more, but are far from offering a roadmap to a sustainable energy security.

There will be significant growth in oil production capacity, especially in Iraq, Brazil, and Canada. But global conventional oil production may have already peaked. It's now up to unconventional resources, whether its Canadian oil sands or Venezuelan heavy oil, or new conventional oil frontiers like deepwater Brazilian subsalt and Arctic resources. It's a lot more expensive to produce, which means oil prices will inevitably rise. But every available oil resource must be put on the market as soon as possible to decrease the chances of another economic meltdown.

There is no silver bullet, and indeed non-OPEC producers, which supply most of the world's oil, are pumping at full capacity, even if "2011 was a year mired with non-OPEC supply shocks. With the exception of the US and Russia, oil production in almost every other major producing country has been much weaker than expected," according to the Merrill Lynch–Bank of America report. Only Saudi Arabia has been able to raise production significantly at short notice, and at a 30-year record[7] in

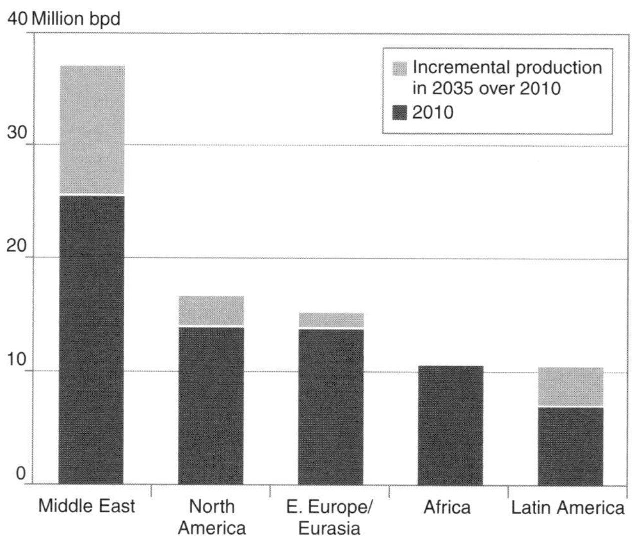

FIGURE 4.4 Oil production in selected regions, 2010 and 2035
Source: IEA WEO 2011

[7]IEA Midterm Energy Outlook June 2011.

order to offset shortfalls elsewhere. Even the ability of Saudi Arabia to significantly increase supplies to the market throughout this decade has come into question. And whatever the truth is about the Kingdom, it's clear that its decisions will be based on its own interests and economic needs, which are balancing its budget, not oil market prices.

There are of course several potential upside surprises. African corruption and political instability, which have weighed down output increases, could be addressed faster than expected; Russia and Mexico might become more welcoming to foreign investors against all odds, or the United States and Iran might become allies and allow Tehran to achieve its potential.

But in a long list of unexpected surprises, Venezuela is perhaps the best placed to significantly improve American energy and national security in the short term. It has the oil, it's technically realistic, and it badly needs to boost production for its own domestic economic and political reasons.

It also has President Hugo Chávez, the anti-American *caudillo* and sworn antagonist of US interests. But unlike other countries, Chávez would be happy to pump more oil if he could, not to help the United States, but to ensure the survival of his regime. The United States has no way to influence him though, and for years it has been losing influence in Latin America.

The United States could, if it chose to, implement policies that will quickly and easily increase oil supplies from Latin America, improving its energy and national security. It could do it without bowing to Chávez or even talking to him, and without sacrificing American values. In fact, the United States could improve its energy security and its geopolitical leverage by simply doing what is in its best interest. But it's not. Its policies are achieving just the opposite. The blind spot is growing into a black hole.

Perhaps another oil shock might be required for Americans to realize that energy is more important to our way of life than the ideological battles that dominate the political scene. And considering global energy supply and demand trends, it might be on the way.

5 AMERICA AND OIL IMPORTS: UNTIL DEATH DO US PART

Crude oil's relation to mankind dates back thousands of years. Asphalt, one of many forms that petroleum can take, was used to build the walls and towers of Babylon. Some even credit oil for saving whales from extinction, although that honor should go to a Polish pharmacist who built the first oil refinery in 1856 and discovered how to distill kerosene from crude oil, a much cheaper alternative to whale oil.

Oil's rise last century to become the world's most important commodity was the result of the internal combustion engine. When people started moving around in vehicles, it was clear that an abundant and versatile fuel was needed, one that would harness lots of energy and that could be easily moved around. Naturally, the military industry led the transition, starting with World War I.

Crude oil is a liquid, easy to transport, and richer in energy content than coal and natural gas. That means it releases more energy than the equivalent amount of other fuels. And more importantly, there is plenty of it spread throughout the world, although heavily concentrated in some regions.

Oil is not only good for gasoline and diesel. There are over a dozen derivatives and hundreds of uses, not all energy related. Oil is pumped out of the ground in multiple forms and qualities. Some oil is heavy, which means it doesn't flow as easily. Some is light, which means it is more fluid. The lighter, the better, although it also depends on how much refining it needs. Crudes can be sweet or sour, which refers to the amount of sulfur in them. Sweeter crudes have less sulfur and are also usually cheaper to process.

Each crude quality yields different kinds and amounts of final products and at different operational costs, depending on the refineries. Modern refineries and petrochemical complexes can yield more products from less crude. Its use for power generation or for heating—which was common until recently—is increasingly limited. And although more than three-quarters of the oil produced is eventually transformed into transport fuels, it is also used to make all sorts of stuff, including plastics, pesticides, fertilizers, wax, and cosmetics.

Yet, despite centuries understanding oil, it remains easy prey for demagoguery, rhetoric, and even lies. That's because numbers are politicians' best friends. Energy data is often manipulated by just about everyone or anything able to generate public opinion. That includes media, nongovernmental organizations, national and international institutions, religious organizations, and so on. Omit a few variables and/or premised information, and you can produce completely diametric conclusions with the same set of statistics. Assumptions are different depending on the agenda, whether it's the end of the fossil fuel era or the beginning of climate change Armageddon.

In 2008, the EIA addressed big changes from one year to the next:[1]

> As policymakers discuss alternative approaches to address high oil prices, demand and supply concerns, participants in the debate often cite estimates of the Nation's dependence on foreign oil. Why is the range of estimates so wide? Like many estimates that reflect multiple variables, the answer depends on "*what*," "*where*," and "*when*" factors.

A case in point, the EIA preview of its 2012 AEO included a massive review of how much oil will be imported into the future. In 2011, it estimated that by 2025 imports would account for 44 percent of total demand, but in 2012 it forecast a significantly lower estimate of 38 percent. That six point difference from one year to the next could hardly be the result of additional domestic production. Indeed, the share of imports is now expected to decrease a lot faster because the economic crisis will dent overall fuel demand more than anticipated. The EIA lowered its estimates of the average miles traveled to reflect "the impact of lower projected economic growth and employment rates." In other words, the best news to American energy security is primarily based on the assumption that Americans will drive less on average because of their decreasing purchasing power.

---

[1]www.eia.gov/oog/info/twip/twiparch/081029/twipprint.html

Even the definitions are subject to manipulation. Proven reserves of any commodity, whether it's coal or copper, are based on the present economic feasibility of production. In other words, proven reserves are revised, not depending on how much oil is underground, but on how much is economically feasible to produce. Higher prices actually help stimulate more production globally, which in turn helps stem price spikes in the future, albeit belatedly.

And then there is politics. Public discourse about energy does not correspond much with reality. For example, Germany's controversial decision to phase out its nuclear power plants by 2022 in the aftermath of the Fukushima meltdown in Japan—especially among friends and neighbors—split society along ideological lines as politicians manipulated public sentiment.[2]

William C. Ramsay, Deputy Executive Director IEA and a former US Deputy Assistant Secretary of State for Energy, wrote then:

> In polarized politics, the voter must take on faith the wisdom inherent in the principled political position. Who cannot be impressed by the nuclear catastrophe at Fukushima, especially after living through the fallout from Chernobyl? But is it reasonable to send the voter to the booth with only half the information? Does the German voter have any idea what it means to shut down its nuclear power in just ten years? Have German politicians made any effort to tell them?[3]

The same criticism applies to the United States as it does to just about any other country. How informed are Americans about energy? Sadly, even less than Germans. That is why populism has led public discourse over the past few decades, not long-term interests. People are either for or against nonfossil alternatives energy sources, as if it is a moral imperative like civil rights or abortion. And it's not. There is no space for ideology.

Yet short-term pump prices are the main drivers of American energy policies. In every election cycle, candidates blast headlines like "American energy independence" and "cutting oil import dependence." Interest

---

[2]Germany approved the phasing out of about 20 gigawatts of nuclear capacity by 2022 in the summer of 2011. Efficiency, natural gas, and renewable energy will offset the loss in the government plan, but experts expect increased natural gas imports from Russia to make up most of the difference.

[3]Ambassador William C. Ramsay in a commentary published in June 2011 by the French Institute of International Relations.

groups and lobbyists, whether it's environmental groups, oil companies, or political radicals, are becoming increasingly involved in the debate.

CSIS's Pumphrey on US energy policies:

> You often have the split in terms of what is working at policy level and what is working at the political level. From political rhetoric point of view energy security is often expressed as how much oil we get and where we import it from. But in the policy sectors there is a clear understanding about global nature of oil market. In many ways the driving factor for energy security is often prices. The concern is over major price disruption and economic dislocations that come after that.
>
> Another illustration of how public discourse and policy-making differ was the so-called "peak-oil" debate in 2008 when prices reached record highs. The world openly wondered whether oil was running out. Well, of course it is. Oil by definition is finite. We just keep extending, year after year, when it will run out. And before that happens, it will become so expensive that other sources will replace it.
>
> Indeed, peak oil is not even an issue since the crisis took hold.

A historical perspective, a future concern

For simplicity, we'll address consumption in the United States of all liquids, including petroleum and all its by-products, biofuels, and other liquid supplies.

Since 1963, supplies of oil and petroleum products have more than doubled. The population has not doubled, although the economy has multiplied by four. In other words, and as explained in the previous chapter, America's economic growth is still hooked on oil but it is increasingly efficient in its consumption.

Excluding all other factors, and looking solely at liquids supplies, regardless of price or origin, domestic demand has sustained two long-lasting reversals. Oil use dipped for around four years as Carter's policies were implemented. Much of that reduction had to do with the transition from oil to natural gas as a feedstock to the petrochemical industry. The second time was during the Great Recession of 2008 that the world is technically still climbing out of. Figure 5.1 shows total US oil and other liquids demand since 1970.

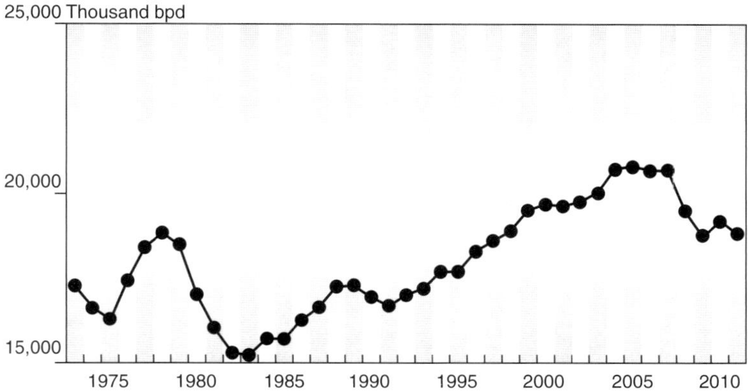

FIGURE 5.1 US oil and other liquids demand

Source: US Energy Information Administration

When political instability involving oil largely subsided in the 1980s and 1990s, once OPEC moved to restrict its political role and Russia was in no place to blackmail other countries, oil flowed endlessly and prices fell. The American economy, despite a few small recessions along the way, grew rapidly. US policymakers bet that American hegemony, coupled with a globalized neoliberal economy, would open up fields everywhere and oil markets would remain well supplied indefinitely. If it's not broken, don't fix it, right?

But policymakers misread oil markets and geopolitics. Oil producers learned the limits of using crude supplies for political purposes. They started investing massively in building up their national oil companies to compete with private Western international oil companies like Exxon, BP, Total, and others. By the new century, most had shed their dependence on Western expertise and giant national oil companies learned to control prices through investment.

At the same time, globalized markets were driving Asian manufacturing and oil demand growth started outpacing supply additions, especially in the twenty-first century. The trend was unstoppable. Non-OPEC oil supply began falling and OPEC gradually closed off access to its fields, all while global oil demand started climbing. The question wasn't about the availability of oil, but about the ability to produce it fast enough to meet rising needs.

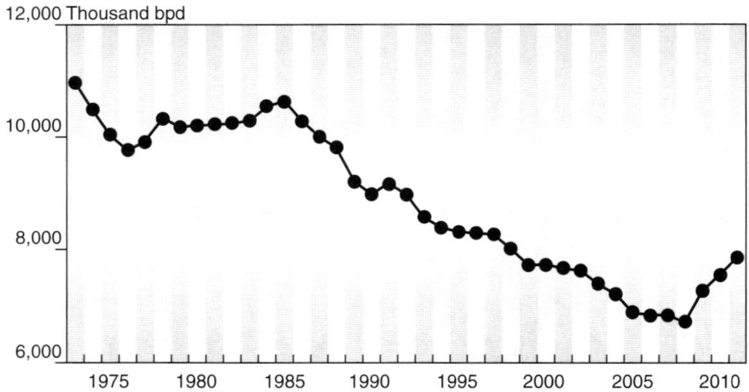

FIGURE 5.2 US oil and other liquids production
Source: US Energy Information Administration

America under the administrations of Bill Clinton and George W. Bush did little to buck the trend. Oil demand continued climbing even as prices started their upward spiral, not over political concerns, but over supply and demand balance. Investment in US oil production was inversely proportional to demand trends. Domestic production was politically unpopular and more expensive in most cases. Offshore drilling and much of Alaska was put off limits. There were little incentives, perhaps under the incorrect assumption that it was best to hold on to it for when it was really needed. The realization came a decade later, approaching 2010, but the infrastructure was nowhere near ready and money for exploration suddenly dried up when the crisis took hold.

It was just easier to go overseas. In fact, American policymakers were especially wrong about Latin America, as we will analyze in Chapter 7. They believed in the 1990s that Venezuela would be able to make up for Middle Eastern supply risks. After all, it had more than enough oil to do so and its plans promised additional production perhaps as large as all of the Saudi Arabian spare capacity. The United States didn't anticipate the rise of Chávez, neopopulism, and reserve nationalism that swept the world, despite multiple warnings.

Eventually, starting this century, oil markets started concluding that we were entering a new era. Global production was not going to pick up with demand. Non-OPEC producers turned to biofuels to offset falling output,

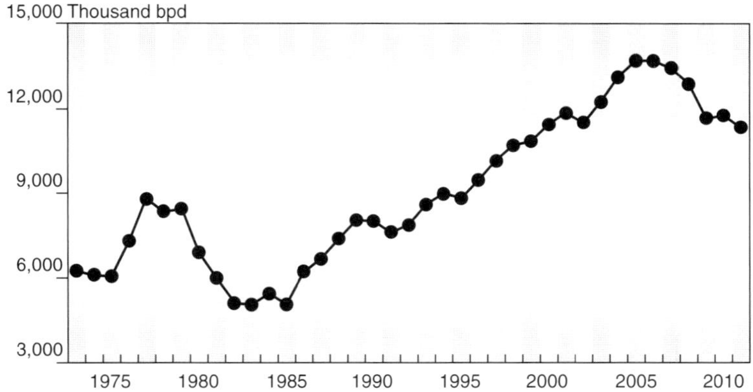

FIGURE 5.3 US imports of oil and other liquids

Source: US Energy Information Administration

while demand picked up in developing nations, the perfect conditions for OPEC to control supply additions and prices for years to come. Iranian production was not going to increase significantly under sanctions. Iraq was not going to be a fountain of oil after Saddam Hussein's death, as many imagined. Venezuela was not going to be anyone's savior as long as Chávez consolidated his power. And that was that.

The US confronted a tightening oil market precisely when it was least ready for it. It's not a coincidence of course. The drop in oil demand when the world went into recession in 2009 was only a temporary reprieve because demand growth was driven by developing countries. It also set the stage for the price hikes in 2011 and 2012, precisely as developed economies face the threat of a second prolonged recession.

Looking into the future, global demand will outstrip supply during most of this decade. In the latest IEA medium term forecast to 2016, geopolitics and increasing demand couple to reduce spare capacity. Prices will be between $15 and $20 per barrel higher through 2016. And in the short term, future additional supplies are likely to be governed by some of the most anti-American regimes, including Iran, Iraq, and Venezuela.

So the bottom line is that simple. The United States needs to be able to secure at least the same amount of liquid supplies for the remainder of the decade and beyond precisely as cheap supplies dwindle and energy prices rise. Even assuming all ambitious programs already in place to trim demand work, the economy will still need cheaper oil prices to avoid a painfully slow recovery that upends all other American foreign policy and national security plans.

Negroponte on supply disruptions versus price concerns:

> There are these risks posed by short term supply disruptions, but there is a self-corrective aspect to that. Between that, the price factor, and our technological sector, I'm not pessimistic about our ability to deal with energy issues. Although I don't like it, I don't think high energy prices are necessarily bad in terms of providing incentives to develop alternatives. The industry also believes fluctuations don't help anybody. Price stability is as important as prices.

Pumphrey on oil supply and price concerns:

> The US will drive up the price if it has to, like the Chinese did in 2004–2005. I don't think that there is any concern that the US won't be able to get oil. The oil will come. There's a concern about the price. It takes time for infrastructure to come to places which can lead to a great deal of volatility. Prices spike for a period of time and then come down.

Whatever the case, high oil prices amid economic stagnation is a perfect recipe for more unemployment, increased deficit, and more uncertainty, a scenario most Americans will not happily accept. It won't matter which party is in the White House.

This is no secret. It's plastered in all sorts of government and private industry reports. This is precisely one of the Obama administration's worst fears. Despite more than a trillion dollars in stimulus funds to avert a total economic meltdown during the recession, and at the expense of the government deficit and fiscal health, another recession could be around the corner.

America's options are just too limited, says Pumphrey.

> You're more likely to see a repeat of 2008, when prices in combination with other factors drove a major economic recession which cut demand, before we see agreement on a full set of policies that would give you a long run solution. The problem is that there is no viable short term solution.

The United States can't do much more about physically raising oil production. Just about all it can do really is diminish geopolitical risks hampering more rapid development. The ultimate question thus for the White House, for Republican candidates, and for every American is whether to wait for the next price hike or do something about it now, something that will have short-term impact and that doesn't betray US values.

But Washington relies on "faith," as Pumphrey described it, that the market is self-regulating, and on top of that, the US government can do little to spur production elsewhere.

Pumphrey on the energy and geopolitics cocktail:

> The most important place where you will see alignment of national security to energy flows is the commitment the US has made to maintain open sea ways, in particular through the Strait of Hormuz. That is very much seen from a policy point of view as a necessary role for the US to play as the single world power. We have capability to project force to do that. So maintaining stability in energy flows is seen in many ways as a global obligation, not necessarily in terms of what is flowing to the US.

Here is Negroponte on the same:

> It's probably not quite as clear cut as it might seem to some people from outside. The factors that influence energy and foreign policy are really very diverse, multi-faceted. It's not about any one single factor. When you think about Middle East energy, the Persian Gulf, energy looms larger because it's such a huge source of energy for the US and the world.
>
> Energy is a factor, but very rarely, if at all, can I think of a case where energy has been the decisive factor in a relationship with a country. The closest you could come to is a country like Saudi Arabia.

Is that the best that Washington has to offer? Oil prices are threatening America's economy and will undermine national security throughout this decade. There is little that can be done domestically about that. Nor can the American government dictate to another country how it should exploit its oil. But clearly there is a myriad of other options available, like with any other pressing issue. It's just a question of resourceful policymaking and willingness.

The Obama blueprint: Rehash or game changer?

US energy security, thus national security, translates today into less oil demand in the United States and more oil production everywhere, including within its own borders. And that's when we return to numbers and politics.

As already discussed, Reagan counterbalanced Carter's regulatory drive into energy and energy fell off the radar screen. Bush "the first" tried unsuccessfully to reinforce the pro-exploration, antiregulatory side defended by Republicans, but a Democratic Congress prevented it; Clinton brought the policy back to center giving alternative energy sources a boost while relaxing exploration slightly, but efforts to reregulate energy did not prosper under a Republican Congress. Bush "the second" reversed policy against regulations and favored fossil fuel development, but with little success in expanding offshore exploration, especially under a Democratic Congress.

So how does the Obama administration fit into the lethargic energy policy-making process in the United States? He came to power following a presidential campaign heavily focused on energy and the economy. Pump prices had just reached record prices and climate change had become a global issue, all while a recession and a historic fiscal deficit crippled policy options.

On March 30, 2011, President Barack Obama gave a speech on American energy security to students at Georgetown University. It was a landmark speech because energy once again became an American priority after three decades of being obscured as a secondary issue.

Obama is betting much of his legacy and reelection, of course, on reinvigorating the energy debate in America, so far unsuccessfully. By equating economic prosperity to an energy revolution, he was hoping his proposals would attract votes disenchanted with the sluggish recovery. But the decisions have backfired and he has backtracked.

The first part of Obama's speech attempted to diagnose where America has gone wrong. There was little word-mincing or manipulation. They were frank, almost admonishing words of America's leadership—which Obama is of course part of—in its inability to improve energy security.

In a matter of months, we've seen regimes toppled. We've seen democracy take root in North Africa and in the Middle East. We've witnessed a terrible earthquake, a catastrophic tsunami, a nuclear emergency that has battered one of our strongest allies and closest friends in the world's third-largest economy. We've led an international effort in Libya.

And it's natural, I think, to feel anxious about what all of this means for us. And one big area of concern has been the cost and security of our energy. Obviously, the situation in the Middle East implicates our energy security. The situation in Japan leads us to ask questions about our energy sources.

In an economy that relies so heavily on oil, rising prices at the pump affect everybody.

Now, here's the thing—we have been down this road before. Remember, it was just three years ago that gas prices topped $4 a gallon. There was a lot of hue and cry, a lot of fulminating and hand-wringing, but nothing actually happened.

The truth is, none of these gimmicks, none of these slogans made a bit of difference. When gas prices finally did fall, it was mostly because the global recession had led to less demand for oil.

The point is the ups and downs in gas prices historically have tended to be temporary. But when you look at the long-term trends, there are going to be more ups in gas prices than downs in gas prices. It is absolutely certain that demand will go up a lot faster than supply. It's just a fact.

So here's the bottom line: There are no quick fixes. Anybody who tells you otherwise isn't telling you the truth. And we will keep on being a victim to shifts in the oil market until we finally get serious about a long-term policy for a secure, affordable energy future.

I talked about reducing America's dependence on oil when I was running for President. But I've got to be honest. We've run into the same political gridlock, the same inertia that has held us back for decades.

That has to change. That has to change. We cannot keep going from shock when gas prices go up to trance when they go back down—we go back to doing the same things we've been doing until the next time there's a price spike, and then we're shocked again. We can't rush to propose action when gas prices are high and then hit the snooze button when they fall again. We can't keep on doing that.

> The United States of America cannot afford to bet our long-term prosperity, our long-term security on a resource that will eventually run out, and even before it runs out will get more and more expensive to extract from the ground.

Obama announced his "reasonable" and "achievable" goal of cutting crude oil imports of 11 million bpd when he took office by one-third in "a little more than a decade" from 2011. It was part of his administration's "Blueprint for a Secure Energy Future." That would imply importing around 3.5 million bpd less crude in about 10 years. The recession did most of the work, as demand slumped with the economy. And now, the goal is within reach, according to the EIA, mostly because Americans will drive less average miles into the future, and because of the significant increase in domestic production of unconventional oil and other liquids, especially biofuels.

US liquids demand, most of which goes into transportation,[4] will increase very little between 2010 and 2025 or beyond, according to the EIA's 2012 preview. In 2011 it had forecast demand rising nearly 10 percent by 2025, or around 2 million bpd, but the crisis triggered a huge downward revision of how much Americans will drive in the future. In addition, domestic liquids production in 2010 of nearly 10 million bpd will increase around 25 percent by 2020, almost two-thirds of which will come from liquids other than crude oil, which includes biofuels.

Furthermore, the EIA's revised forecasts are largely based on decreased fuel demand for transportation, an assumption that has historically fallen short of expectations and that many economists don't trust.

Still, rising oil production in the United States is already playing an important role in improving energy security economically, by creating jobs and allowing most of the billions of dollars generated to stay inside the country.

Giusti on rising production in the United States:

> It looks like "the cavalry is coming to the rescue of the United States." In recent years new technologies of hydraulic fracturing have allowed a wide exploitation of oil and gas in shales. This has reversed forecasts

[4]EIA AEO 2011: Liquid fuel consumption in the transportation sector continues to grow in the Extended Policies case, from 13.6 million bpd in 2009 to 14.1 million in 2025 and 14.2 million in 2035, but at a slower rate than in the Reference case. Cumulative consumption of liquid fuel for transportation between 2017 and 2035 drops by 6.5 billion barrels, or 6 percent, in comparison with the Reference case.

of future natural gas production to the point that very soon the United States will be an exporter of gas. Additional shale gas proven reserves already stand at 860 Tcf and much larger volumes will soon be added. But it is not only natural gas. In southern Texas, North Dakota and Montana, huge accumulations of shale oil have been discovered.

These examples and many other similar ones have been the result of efforts by the private sector. What the new shale oil and gas areas have in common besides their geology is that most mineral rights are in private hands, while federal land holdings are extremely small, where access is burdened by endless federal reviews and restrictions. A study by CERA has determined that shale oil reserves in the US could reach 17 billion barrels. This new access to domestic oil has allowed the US production to increase for the first time in 26 years and projections indicate a possible increase of 2 million bpd by the end of this decade. A study by the National Petroleum Council ordered by Secretary Steven Chu, concludes that oil production in the US could reach 20 million bpd by 2035. Despite the ineffective and hesitant energy policies of the current US administration, a new virtuous wave has emerged and 110 years after Spindletop, the US is again becoming the new frontier. Finally, it is worthwhile to mention the Keystone pipeline, which will eventually be approved and will add more flexibility to oil production and marketing in North America.

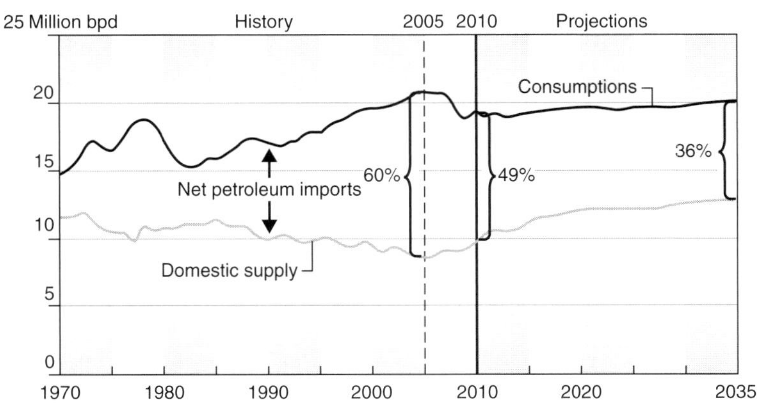

FIGURE 5.4 US oil and other liquids import requirements, 1970–2035

Source: US Energy Information Administration Annual Energy Outlook 2012 early release

"I set this goal knowing that we're still going to have to import some oil. It will remain an important part of our energy portfolio for quite some time, until we've gotten alternative energy strategies fully in force," Obama said in his speech. "The fact of the matter is, is that for quite some time, America is going to be still dependent on oil in making its economy work."

The unexpected BP Gulf of Mexico accident and oil spill had already reinforced the energy debate in 2010, especially after Obama failed at inspiring Congressional support for his hallmark climate change legislation, even among Democrats. That makes Obama's energy roadmap laid out in his Georgetown speech all the more relevant, as it intended to turn a page. The priority was "first, finding and producing more oil at home and second, reducing our overall dependence on oil with cleaner alternative fuels and greater efficiency." The administration early on had signaled its intention to stimulate crude production by streamlining the regulatory process, creating incentives to explore more land that has already been leased out, and opening up more land for exploration, offshore and onshore. In its blueprint it restated this.

But political reality is another. The Deepwater Horizon rig accident, the worst in American history, undeniably set back energy plans in the United States for some years, regardless of responsibilities or the debate over the environmental impact.

Negroponte on Obama's policies:

> It got interrupted. Acts of God and nature happen and people react. Actually, I thought the administration managed it fairly well. They never abandoned their commitment to offshore drilling and they seem willing to see it resume on a carefully regulated basis, but it did get severely interrupted by the oil spill.

Pumphrey went further:

> The administration was in a positive path way before the oil spill in taking tentative step to open up exploration. The oil spill sort of pushed that back and has been put off horizon and that is not a good thing. We should continue to look at areas that we can open up safely. There needs to be further opening up of offshore and politics should not intervene.
>
> We have perpetuated some of our import dependency by exporting where development takes place. We deny our offshore, but import

> offshore oil from Angola and Nigeria, and will eventually from Brazil, and that is an inconsistency that has always bothered me.

Domestic oil production will indeed increase and by 2020 reach 6.7 million bpd, similar to 1994 levels, an accomplishment that can be mostly attributed to high oil prices that have spurred investments into new techniques. But that will have little effect on prices.

Obama's second priority, that is, developing cleaner alternatives to fossil fuels and to become more efficient, will be harder to achieve and contribute even less to decreasing energy prices. Efficiency measures in buildings require a high initial investment and even with additional government funds allocated, the transition will be slow and will barely dent total consumption this decade, especially following the recession. More importantly, it hasn't worked in any country so far, a reality explained through the Jevons Paradox,[5] an economic theory which argues that efficiency only leads to more spending, not less. But even if it had the potential to be a game changer, it's certainly not happening this decade.

Perhaps the most significant efficiency measure that was implemented early on in Obama's administration was increasing the minimum miles per gallon (mpg) for vehicles for the first time in 30 years. Average fuel economy will be raised by 2016 to 35.5 miles to the gallon while Europe is targeting 60 mpg by the same year. Still, Obama said, the measure will save 1.8 billion barrels of oil over the lifetime of covered vehicles.

It's uncertain whether electric vehicles, on the other hand, will play a significant role in reducing fuel consumption, even with the government goal of having 1 million of them by 2015. That's because it remains to be seen what fuel would be used to generate the electricity that the vehicles would be using, which will more than likely be natural gas, a fossil fuel. Obama said the country will invest in renewable electricity, namely wind powered, but the cost of installation and of generation will make renewable electricity a lot more expensive than gas, for example.

Obama said:

> So at moments like these, sacrificing these investments in research and development, in supporting clean energy technologies, that would weaken our energy economy and make us more dependent on

---

[5]The question of coal. 1866. William Stanley Jevons: www.econlib.org/library/YPDBooks/Jevons/jvnCQ0.html

> oil. That's not a game plan to win the future. That's a vision to keep us mired in the past. I will not accept that outcome for the United States of America. We are not going to do that.

But even with a renewable energy vision, there is no viable alternative for oil in the transport sector just yet. Oil hasn't been used in America's power sector for some time, meaning renewable gains don't affect oil prices.

And as for biofuels, excluding the surrounding debate over food security and real fuel savings, the EIA and IEA still expect growth is going to be significant and drive decreases in oil imports. In 2012, the EIA estimated demand will be equivalent to 1 million bpd by 2024. But many believe this target is overly ambitious considering the unsustainable pressure it would create on politically sensitive food prices, not to mention it would only increase US oil demand.

Pumphrey about Obama's policies:

> It's realistic that [Obama's goals] could be met, but I don't know that political will has yet been established to do it. It would take probably stronger efficiency rules, which is the Democratic side of the house; it would take more action to increase domestic production, which is the Republican side, so you need agreement to follow both strategies and it would take some sort of price signal, like gasoline taxes, so I don't think the politics are in place to be able to do it.

Obama said:

> Now, even if we increase domestic oil production, that is not going to be the long-term solution to our energy challenge. Even if we drilled every drop of oil out of every single one of the reserves that we possess—offshore and onshore—it still wouldn't be enough to meet our long-term needs. We consume about 25 percent of the world's oil. We only have 2 percent of the reserves.

Obama's realist approach contrasts his administration's policies. The disappointment is a flashback of Bill Clinton's administration. In the same way that it took anticommunist Nixon to open up China, in the topsy-turvy ideological divide on energy and environment, it would take a Democrat to level with the electorate on energy reality. Obama, who set his own bar for improving American energy security, is missing this opportunity.

Obama has failed to offer any practical solution to energy security simply because, like others before him, his policies are politically driven and fail to address the long-term priorities. Decreasing oil imports should be part of a broader strategy—not the main goal—that ultimately addresses the unsustainably high energy prices and market volatility undermining America's economic growth and ability to exert power on a global scale. More than ever, a new approach is needed, especially because America's biggest rivals are either big oil producers, like Iran and Russia, or big oil consumers that will compete for insufficient resources, like China and India. Indeed, the Obama blueprint, regardless of how useful it might be on paper, does not address oil market volatility or unsustainably high prices, and will almost certainly fall into the same bipartisan quid pro quo.

More importantly, reducing American imports will not translate into more energy security because the biggest threat is in the form of higher sustained prices, and there is little net change to the global oil balance or to the tight market as a result of US shifting energy trends.

After all, it's not where America's energy supplies come from, but at what price.

6 GLOBAL OIL SUPPLIES AND AMERICAN SECURITY

Up to the mid-1970s, the United States was the world's biggest oil producer, although it started importing to compensate for insufficient output in the 1950s. America's imports, as addressed in the previous chapters, increased up to 2005 with only a few hiccups. Prices were only an issue in case of geopolitical disruptions. Capitalism, the theory goes, would make sure everyone was supplied. More demand would inevitably bring more supplies. And indeed that is exactly what happened last century, the century of oil consumers.

A future in which oil supplies are more expensive and harder to come by, one in which it gets harder and harder to keep up with demand growth, was practically unimaginable. But starting in the 1990s, many heavily populated nations like China, India, Brazil, Saudi Arabia, Iran, and Turkey started expanding their economies rapidly and simultaneously. Most of the easy-to-access oil in the United States and Europe had already peaked. Costs started rising significantly, not just because of demand, but because increasingly the oil coming into the market was more expensive to produce. Furthermore oil was increasingly managed by governments and their national oil companies with more than just profit in mind.

The world changed and few noticed. Oil production, markets, and prices increasingly began being shaped by Russia, OPEC, and harder-to-produce petroleum, instead of consumers. As the world transitioned into a new century, the era of oil consumers was replaced by the era of oil producers.

The good news is that there are enormous amounts of oil in all its forms and we have only really started scratching the surface. But having oil underground doesn't mean it's coming out. Even oil fields under production will have different recovery rates that vary between 10 percent and 80 percent. And recoverable, as explained before, is defined by current market conditions and available technology.

Let's just say the world has between 4.5 trillion barrels and 5 trillion barrels of oil,[1] which really just sounds like a ridiculously large number. About a third of that is yet to be discovered. To put reserves into context, current global consumption of about 30 billion barrels a year is about 0.5 percent of the world's known global oil supplies annually. That will change little even with increased demand.

In fact, oil with a 90 percent probability of being pumped out, also known as proven reserves, stands currently at a whopping 1.4 trillion barrels, of which we consume around 2 percent annually. But proven reserves have doubled since 1980 and increased one-third since 2000. Those upward revisions are likely to continue, although not necessarily as fast as demand climbs. Still, that is oil we know will make it to motors someday.

Most of the world's oil is concentrated in the Middle East, South America, Canada, Russia, northern Africa, and around the Caspian. Countries often claim the world's biggest oil reserves. If we include some of the harder-to-extract low quality crude, Venezuela is the top reserve holder, followed by Saudi Arabia, Canada, Iran, and Iraq. Barring a lingering civil war, there is a good chance that by the end of the decade Iraq might be able to displace one or two of the reigning oil richest countries.

In terms of quality, Saudis, Iranians, and Iraqis sit on huge reserves of good quality oil, or conventional, while most of what Venezuela and Canada hold is harder-to-extract viscous oil, or unconventional. When it comes to production, Saudi Arabia is and will remain king not just through 2020, but beyond, simply because most of the infrastructure to produce and export oil has already been laid. Venezuela and Canada reserves will take a lot longer to develop and it's still uncertain how much of it will be pumped.

The pressing questions are how much can be produced, how much can be exported, how much will be consumed, all within a given time frame. And, as premised in earlier chapters, the United States—indeed the world—is entering an especially uncertain decade when it comes to oil and energy security. What will oil supply and demand look like through 2020? Where will it come from? And at what price will it be sold?

---

[1]Based on assumptions used by the IEA, split into 2.5 trillion of conventional oil estimated by the US Geological Survey, and between 2 and 3 trillion of unconventional.

Oil 101

Conventional oil is what most people imagine when they think of petroleum, that is, rigs pumping oil out of a deep hole. It also includes offshore oil, even if it's a lot more expensive to extract than the onshore variety. In fact, some conventional oil can be just as difficult and costly to extract as unconventional. That is the case of Brazil's recently found huge oil deposits known as the presalt, which lies beneath 1.25 miles of salt, itself below 1.25 miles of rock, itself below 1.25 miles of water. It's the new oil frontier.

Much more conventional oil awaits to be discovered and extracted, especially in the Arctic and the presalt layers off Africa, but these will not be contributing much, if at all, this decade. Brazil's development started years back. Still, there have been several delays and more are expected, especially with rising concern over the safety of offshore drilling, but we will analyze that further in Chapter 7.

A big part, perhaps most, of the onshore conventional sources has been long discovered and for the most part has been producing for some time. And many of those fields, especially those in industrialized nations, have peaked. That means the output began falling after a long plateau of maximum production and will continue falling until it becomes economically nonviable to produce.

The IEA estimates around a third of the existing conventional oil has been already pumped out, and by 2035 as much as half of it will, 61 percent of it in non-OPEC countries and 47 percent in OPEC countries. The contribution of technology is critical in this regard. Countries have been able to reverse declining trends temporarily, or simply sustain output levels, thanks to new and constantly improving oil recovery techniques. In the case of the United States, production is reversing decades of losses, and it's mostly thanks to enhanced oil recovery.[2]

---

[2]According to the IEA WEO 2010 report: "Almost half of the increase in proven reserves in recent years has come from revisions to estimates of reserves in fields already in production, rather than new discoveries. Although discoveries have picked up in recent years with increased exploration activity (prompted by higher oil prices), they continue to lag production by a considerable margin: in 2000–2009, discoveries replaced only one out of every two barrels produced—slightly less than in the 1990s (even though the amount of oil found increased marginally)—the reverse of what happened in the 1960s and 1970s, when discoveries far exceeded production. The contribution of offshore discoveries, including deepwater, has increased significantly since the early 1990s. Since 2000, more than half of all the oil that has been discovered is in deep water. Although some giant fields have been found, the average size of fields being discovered has continued to fall. The New Policies Scenario requires average annual development of 9 billion barrels of new discoveries from 2015 onwards."

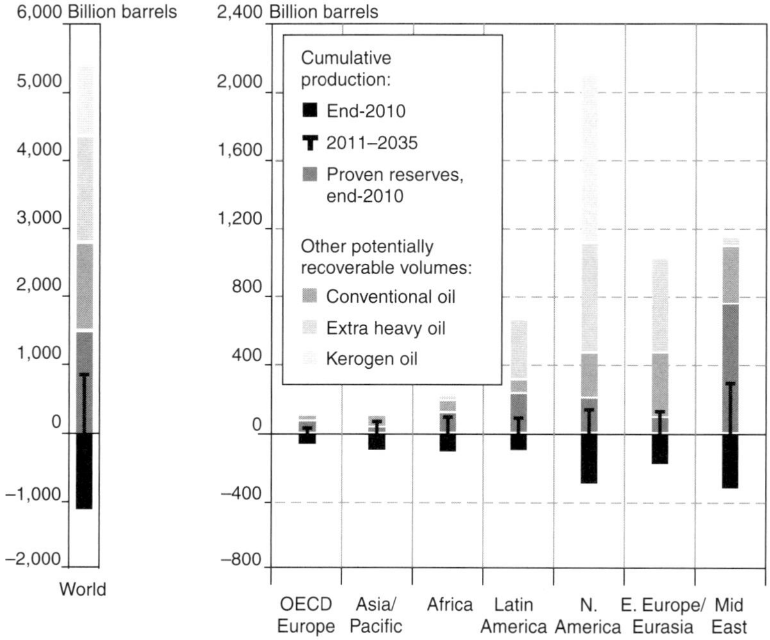

FIGURE 6.1 Recoverable oil resources and production by region and type
Source: IEA WEO 2011

Unconventional sources on the other hand are plentiful, perhaps three times larger than conventional ones. They have only recently come into play though, and a lot more technology deployment is required to understand just how much of it will make it out of the ground, that is, how much is recoverable.

Suffice to say that most of the increase in proven reserves over the past two decades has come from Venezuelan heavy oil and Canadian oil sands, both unconventional sources. Another source, which is increasingly being developed in the United States, is tight oil, an unconventional source produced with similar techniques as those used for shale gas that is found in fine-grained sedimentary rock. "Onshore tight oil production accounts for 31 percent of lower 48 onshore oil production in 2035, compared with 12 percent in 2010," said the EIA in its 2012 outlook.

Unconventional oil though requires a lot more processing and money to get out of the ground and into some consumer-friendly form. It often

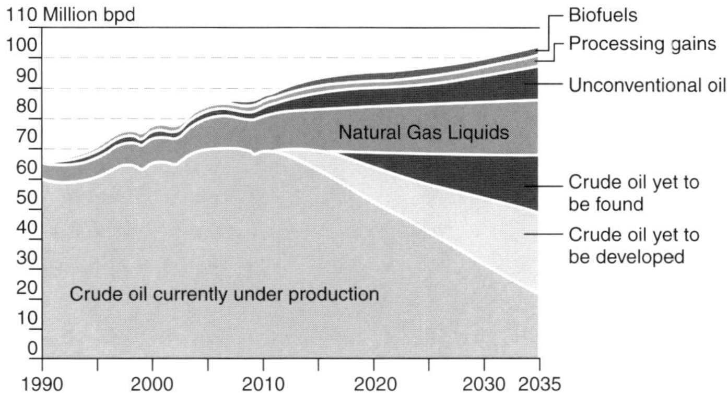

FIGURE 6.2 World liquids supply by type

Source: IEA WEO 2011

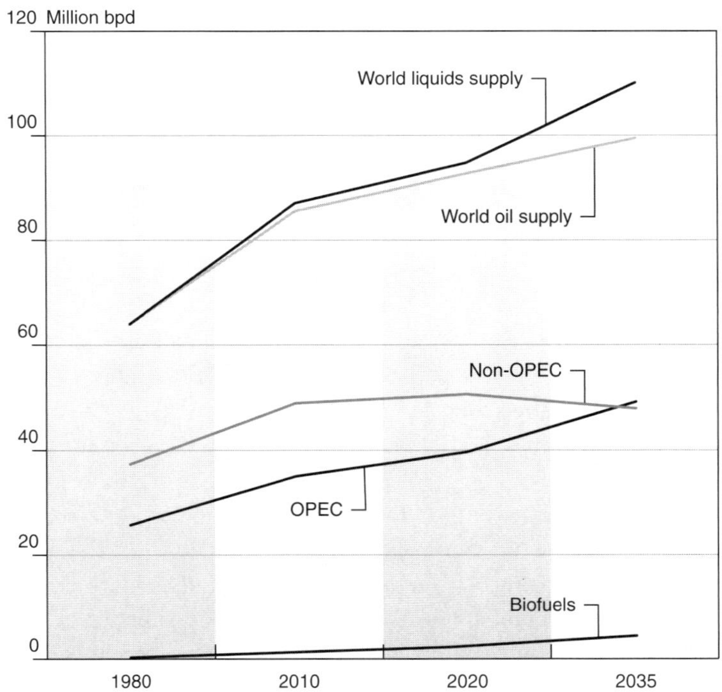

FIGURE 6.3 Oil production and supply by source

Source: IEA WEO 2011

has to be liquefied before separating, usually by heating up and adding solvents. Canadian oil sands, also known as bitumen, are more like asphalt coming out of the ground. The costs of extracting, delivering to refineries, and refining are higher, although these are likely to fall gradually with new technology.

Unconventional oil is also less environmentally friendly than its conventional cousins. Its production has a greater impact on natural habitats and landscapes, but more importantly, it releases significantly more carbon dioxide, the gas blamed for global warming. It doesn't matter if climate change is real or not. It raises costs. Governments are still figuring out how to best develop these resources. Global liquid supply by 2020 is expected to grow a staggering 9 percent from 2010 to around 95 million bpd, according to American and OECD oil market estimates, including conventional crude oil, natural gas liquids, unconventional oil, processing gains from improved efficiency in refineries, and biofuels.[3] Slightly more than half of that will come from non-OPEC countries.

OPEC countries will contribute the most to that growth with a 20 percent boost to 40.5 million bpd, while non-OPEC countries pumping at maximum capacity will increase supplies by 10 percent to 48.2 million bpd. The rest comes from processing gains and biofuels.

The target is ambitious, especially amid a slow economic recovery and increased uncertainty over the possibility of a double-dip recession. On top of that, heightened investment risks in the MENA region amid the evolving Arab Spring are undermining OPEC production additions plans, all while internal demands skyrocket.

There is variable optimism for Iraq, Canada, Brazil, Angola, the United Arab Emirates, the United States, Colombia, and biofuels. In contrast, there is little hope for Venezuela, the best positioned on paper.

Most of the expected gains will go to meet the fuel demand in developing economies. Middle Eastern exports, which account for about half of the total traded crude, will basically go to satisfy the more lucrative Asian demand and its own soaring internal consumption to meet power and desalination plant requirements.

Markets will become even more global and liquid led by African and Middle Eastern shipments to Asia, while long-haul supplies to OECD countries will decrease. The United States is expected to continue decreasing imports from the Middle East and Africa, and to increase

[3]The IEA forecast is a little smaller than the US EIA's, but the trends are similar.

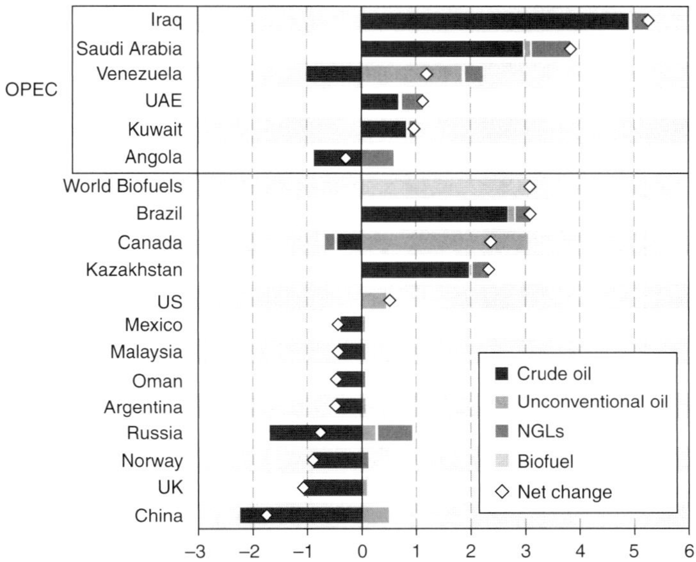

FIGURE 6.4 Major changes in liquids supply, 2010–2035

Source: IEA WEO 2011

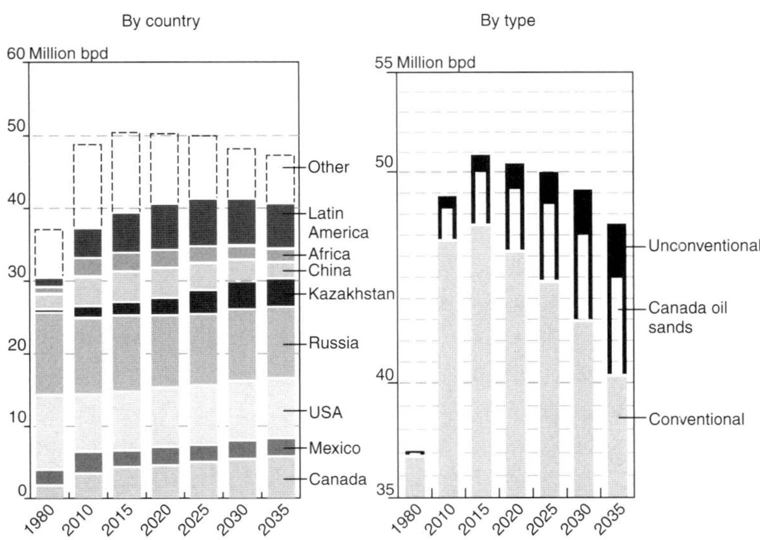

FIGURE 6.5 Non-OPEC liquids production by country and type, 1980–2035 (includes crude, NGLs, and unconventional)

Source: IEA WEO 2011

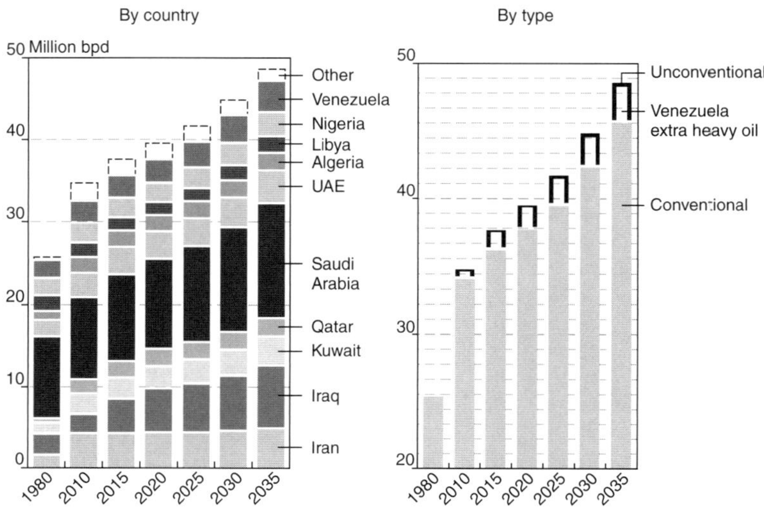

FIGURE 6.6 OPEC liquids production by country and type, 1980–2035 (includes crude, NGLs, and unconventional)

Source: IEA WEO 2011

supplies from Canada and Latin America. The IEA expects additional production in Latin America will go to meet rising internal demand and to export to Asian markets, namely China.

The tight market and prices

Markets interpret reported figures, forecasts, and dozens of assessments to calculate current and future fluctuations in supply, demand, and costs. And in turn, these determine oil prices, the ultimate concern of consumers and governments alike. There are several benchmarks—depending on origin—to price the different crude qualities out there. The United States has one benchmark, Europe has another, Asia signs long-term supply deals, and the Middle East is trying to introduce a third benchmark. WTI stands for West Texas Intermediate and is the standard in the American continent; Brent comes from the North Sea off Britain and is the price reference for Europe and to a lesser extent the Middle East, and the

Dubai-Oman grade is increasingly being used to benchmark Persian Gulf crudes. For this book, the focus remains on WTI.

Until a few years ago, it used to be that world market prices depended almost entirely on the supply and demand balance, also known as *fundamentals*, even during politically driven disruptions like the Iran-Iraq War in the 1980s. Traders worried then about short-term disruptions and the ability to offset them, more than long-term depletion. But as the difference between supply and demand continues to thin out, the market preoccupies itself with geopolitical concerns that can have serious long-term consequences, like political instability that delays oil development projects. Financial markets are also playing a direct role as non-oil traders hedge other investments on more secure assets, such as oil. And as the US deficit weakens the dollar, oil prices rise to offset the currency's losses.

Monitoring even a fraction of these factors is a job for officials and companies, not for consumers. It's hard enough to keep track of pump or flight prices. A good barometer to follow though is the world's spare capacity. It's surplus production capacity: potential minus reality, or spare capacity for short.

It can also be split into nominal and effective spare capacity. The latter, as described by the IEA, tries to account for political and managerial decisions that decrease production, as opposed to crude availability or technical issues.[4] Iraq, Venezuela, and Nigeria illustrate the difference. Non-OPEC countries don't hold spare capacity. Only OPEC countries do.

More often than not, the problem is not technical, but political or managerial. And internal politics, instability, corruption, and mismanagement threaten even those capacity additions that are planned.

In fact, the difference between nominal and effective spare capacity for the next five years is about 1 million bpd, or 25 percent, mostly stemming from Iraq, Nigeria, and Venezuela, which for political and managerial decisions can't physically use some of their nominal spare capacity.

In June 2011, the IEA issued a concerning outlook in its Midterm Energy Outlook:

---

[4]OPEC's "effective" spare capacity recognizes that over the last decade, and on a consistent basis, around 1 million bpd of nominal spare capacity in countries including Iraq, Nigeria, and Venezuela, has not been immediately available to the market for technical, security related, or infrastructure reasons.

By and large, a growing number of OPEC producers appear incapable of increasing production capacity in the medium term due to a multitude of reasons, including:

- uncompetitive contract terms and/or shifting fiscal regimes for international oil companies stemming from resource nationalism (Algeria, Libya, Nigeria, Iran, Venezuela, Ecuador);
- project delays or cancellations due to internal and external political and security constraints (Iran, Nigeria, Libya, Kuwait, Iraq, Ecuador, and Venezuela);
- delays to project developments due to requirements on local sourcing of infrastructure, equipment, and workforce (Algeria, Angola, Nigeria, Libya, Iraq, Venezuela, Ecuador);
- inability to monetize assets in the short term by surging production due to management issues/equipment/expertise constraints (Algeria, Nigeria, Libya, Kuwait, Venezuela, Ecuador); and
- lack of expertise or access to advanced technology for Enhanced Oil Recovery (EOR) projects needed for maximizing recovery rates (Algeria, Libya, Iran, Kuwait, Venezuela, Ecuador).

Notice Venezuela fits into all categories.

Spare capacity is also one of the main gauges for oil traders, oil companies, refineries, and investors when deciding whether to invest billions of dollars in developing a field. If the world doesn't need it, then the oil will be worth less, and in that case it's best left untouched. Those decisions are assessed over time of course. What matters is the price over the long haul, over a decade at least. Many oil companies in 2007, private and state-owned, that faced plummeting prices and complete uncertainty over how long the recession was going to be, decided to delay investment plans, for example.

There was no doubt that demand would pick up, but nobody could be certain when. The economy unexpectedly recovered swiftly, led by China and the rest of Asia and Latin America, which were better able to weather the crisis. It didn't matter if developed economies didn't want as much oil. Growth is driven by the combined weight of developing economies.

"There should be little doubt at this point to anyone that follows oil fundamentals that supply remains exceptionally scarce. We had been looking for a substantial inventory buildup [in the first half of 2012] on the expectation of higher supplies and a decelerating demand picture. If demand for oil is falling sharply all around the world as net oil importers can no longer afford it, why aren't prices coming down as we projected last December?," a Merrill Lynch–Bank of America report asked in 2012. "In part, inventories have failed to build despite the demand declines. Looking across the OECD, we find that inventories remain at the lowest seasonal level since 2007, just a few months before prices spiked to $147/bbl."

Non-OPEC production continues to disappoint, while OPEC spare capacity continues to rapidly shrink. In the first half of 2012 it hovered around similar levels as those in 2007, ahead of the last run-up in prices to nearly $150 a barrel. "It is no coincidence that the last time effective spare capacity headed south of 3 million bpd, prices too were on a sustained rise. Temperatures surrounding the Iranian issue may cool, and non-OPEC output *should* recover as 2012 progresses. Until then, the market's relatively slim 'buffer' suggests a bumpy ride in the months ahead," the IEA said in its March 2012 Oil Report. In May 2012, when oil prices had fallen from seasonal spring highs, and as the market readied for another summer spike, the IEA's chief economist Fatih Birol warned of the dangerous juncture: "Even current prices are far too high for the current economic context . . . and pose a serious risk. I'm concerned for Europe and I'm also very concerned that these high prices would hit the still hesitant and slow US economic recovery."

Despite Libyan production coming online sooner than expected and record Saudi Arabia production, the market is gearing for a thin margin through the middle of this decade, with utter uncertainty of how geopolitics, especially in the Middle East, will affect oil output thereafter. That means prices will remain high in the short term. But it also means that companies will invest more money in coming years to build more production capacity. That is the logic of the market.

OPEC will invest depending on its interests, which means it will keep prices as high as the global economy can conceivably sustain. Most of the spare capacity is and will remain in Saudi Arabia, despite significant production increases especially in the Persian Gulf as additional production is consumed entirely by Asian rising demand. There is growing uncertainty though about OPEC spare capacity, especially because during the Libyan shut-in, only Angola and Saudi Arabia managed to significantly increase output.

The Saudi case illustrates growing concern among consuming nations like the United States. Saudi Arabia is not planning significant production capacity increases through 2016 after having completed hefty investments in 2010, although it says it could easily add another 2.5 million bpd within this decade to reach 15 million bpd. That uncertainty alone is huge. Internal Saudi oil demand is also rising faster than expected, partially due to energy subsidies, but increasingly because of its lack of alternatives to generate electricity. That is, Saudi Arabia for political and economic reasons will seek to maintain its spare capacity, but it will do so to manage prices to its benefit, not to benefit consumers like the United States.

Indeed, in January 2012, Saudi Arabia set its new benchmark for Brent oil at $100, below current prices, but still significantly higher than its previous $75 a barrel mark from 2009. Rising global demand and output constraints cut spare production capacity in 2011 and will do so again in 2012, only to rise slightly through 2016, according to the IEA Midterm Energy Outlook 2011. The agency described it as "a fairly thin supply cushion."

The EIA said in its 2012 preview: "Uncertainty regarding OPEC members' actual investment and production decisions and the degree to which non-OPEC countries and countries outside the OECD restrict access to potentially productive resources contributes to world oil price uncertainty."

According to the Bank of America–Merrill Lynch midterm oil forecast published in February 2012:

> Spare capacity will not change much over the period. Based on our analysis of the IEA's field-by-field database, we find that non-OPEC production is now declining at an average rate of 4.2 percent. However, while progressive decline rates are supporting non-OPEC supply growth, OPEC faces the opposite problem. OPEC decline rates continue to rise, with the IEA estimating that the existing production base is declining at a rate of 4.5% per year, compared to 3.9% observed a year ago, and 3.3% the year prior. Overall, given improving non-OPEC decline rates combined with deteriorating OPEC rates, we estimate that global production is declining at an average annual rate of 4.3%, compared to 4.2% last year.

Global spare capacity this decade will thus remain strained even while the oil industry runs at full steam. Even with huge additions in the Middle East, Canada, and Brazil, it will remain tight. Even if Russian production remains flat at current post-Soviet record levels, even with Caspian and African exports increasing, and without considering unexpected minor or major delays as a result of politics, there is little room for error. The fact that most of the global output growth for the remainder of the decade is expected from Canadian oil sands, Brazilian and Angolan deepwater offshore, American tight oil, and Venezuelan heavy oil, all of which are unconventional or cost significantly more to produce, augur a more bullish than bearish future for oil prices.

The optimistic range, one that doesn't take into account politics, will hover between $100 and $130 a barrel. And the American economy will be hard pressed, if not unable, to assume the higher end over an extended period. Rising budgetary needs in OPEC countries, as well as higher production costs in non-OPEC countries, mostly for unconventional oil, have set an $80 floor on oil prices in the short term, according to the Bank of America–Merrill Lynch midterm oil forecast:

> In a supply-constrained world, however, oil is unlikely to spend much time hovering around its price floor. Rather, we believe oil prices will likely remain a key constraint on global economic growth. This suggests that prices will continue to spike over the next five years to keep on rationing demand back down to the limited available supplies. The world economy can hardly afford to spend more than 9 percent of its GDP on energy. On our estimates, occasional demand rationing episodes could result in prices occasionally spiking to $200/bbl over the next five years. We believe that oil will continue to trade at a substantial premium over other fuels through the end of the decade.

Furthermore, as the Libyan war has exposed, the world is no longer sufficiently supplied to sustain man-made supply disruption like war, nor will it be for the remainder of the decade. The unprecedented release of oil supplies from the developed nations' emergency storage in 2011, led by the United States as the Libyan war raged, highlights just how sensitive oil markets are, and how concerned governments remain. The fact that the United States and its allies are considering once again turning to emergency storage in the summer of 2012 to offset market volatility as an EU embargo of Iranian oil takes hold illustrates the limited options available to Western countries at this juncture. Emergency storage was conceived as a last resort in cases of supply disruption, like the Iraq War and the Katrina Hurricane that shut down Gulf of Mexico production.

But the Obama administration has opted to use America's emergency reserves as part of its geopolitical agenda, specifically to target Iran. OPEC countries have harshly criticized this, not because they are defending Iran, but because the sanctity of market fundamentals will increasingly fall prey to politics if Western countries start manipulating the free-flow of oil for political purposes.

Energy security for the United States requires more spare capacity, especially as geopolitical risks around Iran increase. There's no way around it. Otherwise a double-dip recession—or at least an extended period of slow growth and high unemployment—triggered by sustained high oil prices seems inevitable, one that the world might not be all that ready to sustain.

American military and economic muscle will be further eroded precisely when it's needed the most, as the world reorganizes into multipolar spheres competing increasingly for resources and influence.

The Chinese factor

The single most important driver of shifting dynamics in world oil markets is China. It has a voracious energy appetite, much smaller than America's, but growing so fast that it alone will continue to account for most of the world demand growth throughout this decade and probably the next. At the same time, for several years it has been directly and indirectly responsible for driving global production gains.

This decade will also define a new world order in which new economies and alliances will emerge. Countries like China will simply pay for having a bigger say in world affairs. And the United States needs economic flexibility to navigate this decade.

Both China and the United States need to pursue a similar energy security strategy. The United States has a commanding lead, but China is catching up fast. The two will face this decade under different circumstances. The United States owes nearly $15 trillion, nearly 100 percent of its GDP. In contrast, China's public debt was less than 20 percent of its GDP in 2010 and it has $3 trillion in foreign reserves, a world record that swells every year. The United States has had a trade deficit for decades, while China has annual trade surpluses. America's economy will struggle to grow and China will continue expanding rapidly. The American public deficit is increasing, while the Chinese one is decreasing.

Here's Negroponte on China's inevitable new role:

> I don't know what we're going to do about it. They will do what they want to do and I don't think we can prevent it. People say 'You used to be the most powerful country in the world. What's happened to you?' Well I have news for you. There's another country coming after 150 years operating way below its potential, humiliated by Western colonization, war, occupation, communism, revolution, Cultural Revolution. And it goes on and it goes on for 150 years. It's now actually starting to realize its potential. We're gonna have to treat China as a peer and we are going to need to have a good dialogue with China going forward.
>
> They're looking for some of the same stuff we are. Oil, iron ore. And I seem to remember us doing the same. Isn't that what we did in Latin America with mining companies?

This is not to say China and the United States will literally fight for oil any time soon. There will be no need to. Besides, China is far from challenging the United States militarily. But it should still plan for increased competition. With decades of advantage, American requirements to improve energy security are smaller than those of China, but then again China is in a better position to advance its interests this decade than the United States.

In 2000, China consumed half as much primary energy as the United States. In 2009, it surpassed America as the biggest consumer, according to the IEA. And by 2020, China's demand will increase another 50 percent, while that of the United States and Europe will remain mostly flat.

In terms of oil, China has consumed more than a third of the world's total production additions over the past three decades. It is already the second biggest oil importer after the United States. And its demand will increase around 20 percent more, or 2 million bpd, between 2011 and 2020 to nearly 12 million bpd.[5]

Oil imports already contribute more than half of total demand and Chinese production is expected to remain flat at around 4.5 million bpd. By 2020, more than 60 percent of its oil demand will have to be imported, around three-quarters by 2035. And, yet China will still import less oil than the United States as far as 2025. And unlike the United States, the Chinese import portfolio is well diversified. Its top suppliers are Saudi Arabia and Angola, each with around equal shares that combined still

---

[5]IEA, EIA, and other projections.

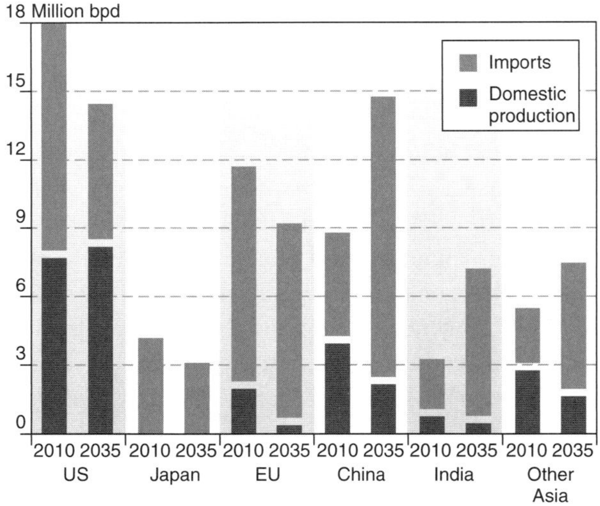

FIGURE 6.7 Oil demand and the share of imports by region, 2010–2035
Source: IEA WEO 2011

supply only a third of total imports. After that come Iran, Oman, and dozens of smaller suppliers, many in Africa and increasingly in Latin America, but China is not heavily exposed like the United States, which imports 70 percent of its oil from the same five countries.

Again, to put those numbers into context, China needs to work very hard to secure enough oil to fuel its economy. Most global demand growth, as mentioned above, will be Chinese, and as such, it will drive global prices. And in the process China will continue securing supplies through oil for cash development projects and investments, a model it has successfully implemented for decades.

China's diversification strategy is to limit its regional dependence on the Middle East, and it has aggressively used its reserves to offer billions in development credit, underwritten with oil, especially in Africa, Asia, Latin America, and even Russia. We will take a closer look at this issue in coming chapters, especially in regard to Latin America, but suffice to say that Chinese oil demand is already a challenge to US energy security, not in terms of securing volumes, but in terms of prices.

As much as China has driven demand, it has also justified huge investments in production throughout the world directly and indirectly. Even when its companies are not involved, China's demand will continue driving investments throughout this decade and beyond. The United States benefits indirectly when prices decrease. In the short term that is unlikely, but eventually global production gains are in America's interests, even if they are ultimately to meet China's appetite.

Here is Pumphrey on Chinese oil trends:

> From energy security point of view, one of the biggest threats that you have to maintaining a stable price in the long run has been satisfying growth in Chinese demand. That's what putting pressure on prices. It's helpful to the extent that Chinese companies are investing in additional production capacity. On balance, it's zero-sum if they compete but still bring oil to the market place. And you have a net positive in oil flow if China invests in developing deposits that aren't attractive to international investors for different profit incentives, different needs, different capital costs.

Further tantalizing in terms of geopolitics is how China moves militarily to secure its oil supplies, much like the United States did back in the 1970s when its own economy was roaring. Broadly speaking, China has little ability to challenge the United States for supremacy, certainly not militarily in Latin America, Africa, or even Asia Pacific, but it could be emboldened closer to home. That is in the long term and not an issue in this book, but the trend is nonetheless inevitable. Sooner or later, if things don't change, the United States and China could scale up their competition for oil resources.

Pumphrey on geopolitical concern over Chinese oil hunts:

> There's always going to be a competition. The bigger concern is at the geopolitical level, which is a question of influence. With US and China, everyone is vaguely uneasy, but not sure why. A lot of the concern is the influence game. What do Chinese think they are buying when they go into a country? Is it just business or influence? And how do you separate that? Does the greater game that emerges between China and US get affected by that? That's the one to watch.

US oil import alternatives

The foreign source of supply is important, but not definitive in terms of energy security. After all, oil can be bought openly in the market. But it's worth looking at the origin of American imports because they will increasingly become more regionalized.

Currently a quarter of total imports come from Canada, a share that is sure to increase into the future. Mexico, Saudi Arabia, Venezuela, and Nigeria will likely see its total sales to the United States drop in line with shrinking imports. All other suppliers are significantly smaller, and of those perhaps only Iraq, Brazil, and Africa could—if they wanted to—export significantly more oil to the United States, although that will depend on how oil markets evolve over the decade.

What appears to be an irreversible trend is that the United States will increase its reliance on continental supplies to meet its oil demand, mostly from Canada, Brazil, and Colombia, the IEA and EIA expect. Venezuelan imports are expected to decrease due to bilateral relations. In contrast, Mexican production and exports will continue a downward trend.

The top five sources of oil imports are unlikely to change significantly. It's not that the United States couldn't replace around 1 million bpd of Venezuelan total petroleum products, for example, with other sources. It could if refiners find a better supplier, like Canada of course. But it would be irrelevant to oil prices because the oil would just be bought by another country, like China.

Canada, of course, is America's favorite partner. Canada's ability to produce a lot more beyond what it's already expected to is limited though. Only Iraq, with its very particular circumstances, and Brazil are expected to add more.

Latin America, which has the world's second biggest oil reserves, offers the best chance of adding significantly more oil output on top of current plans. It's that simple. Latin America's potential is seriously underestimated due to political concerns. An upside surprise though, considering its huge potential, would inevitably drive down prices.

Canada, risk, and promise

Canada is already America's closest ally in just about every respect, including energy. It will add at least another 1 million bpd of net supply growth by 2020, the third biggest contribution in that period, behind Iraq and Brazil. America's energy supply will grow increasingly secure largely thanks to Canada's oil development plans.

The energy relation is a merger of interests. Canada has a lot of oil to sell into the future that is already aligned with a big part of American refining capacity. The United States already buys most of Canada's available oil exports and will remain the undisputed top oil consumer for years.

The risk, in fact, is that a lot is expected of Canada, perhaps too much. Most of the country's oil wealth comes from oil sands, also known as tar sands. They are more environmentally damaging, more expensive to produce, and have a bigger carbon footprint than conventional oil, and as such face significant headwind and likely delays.

Canada has the world's third biggest reserves, after Venezuela and Saudi Arabia. It has 175 billion barrels, three-quarters of which are in the form of oil sands in Alberta, close to the northwestern border, across Montana. It already produces nearly 3.5 million bpd, of which it exports anything in excess of its 2.3 million bpd demand. By 2020, the IEA expects 4.5 million bpd, with growth from oil sands offsetting depletion in conventional wells, according to the IEA's 2011 WEO. The increased supply will go almost entirely to world oil markets, but its destination is still in doubt.

Around 45 percent of the supply already comes from oil sands, a share that will increase to 65 percent by 2020 and nearly 80 percent by 2035, according to the IEA. Oil sands are landlocked and increased supply needs more pipeline infrastructure. The only options are south to the United States or west to China.

A big part of the problem is that the United States will significantly decrease the amount of oil it imports going forward, which means that the additional Canadian production could at best replace imports from other countries. The United States might be well served politically to burn more Canadian crude and decrease purchases from other clients like Venezuela, some African regimes, and perhaps even in the Persian Gulf. But that is not a government choice, but a private industry one.

In any case, in January 2012 the Obama administration unexpectedly denied approval of a 500,000 bpd pipeline expansion known as Keystone XL that would connect oil sands to refinery hubs in Oklahoma in the first leg, and all the way down to export hubs in Texas in the second leg. Its approval got caught in a bipartisan dispute that had little to do with energy, but that nonetheless could have a profound effect on global oil markets.

The proposed route crossed over a water aquifer, and many Republicans joined Democrats to demand a less risky alternative. Environmentalists also pressured the White House over climate change concerns, and faced with an increasingly bitter debate, Obama decided to take the heat and deny the permit temporarily until a new route is proposed.

Pumphrey on the Keystone XL:

> One of the Obama administration's strongest political bases is opposing it, but from energy point of view it's good policy. So how can you have both things?

Negroponte criticizes the Obama administration's reluctance:

> We can't even get a pipeline approved to carry more oil between Canada and the United States. Now that's a blind spot.

Streamlining approval of more pipeline interconnection between both countries will increase investor interest and accelerate oil sand development. The ultimate decision is now delayed until 2013, but Canada is not waiting in the meantime. "I am very serious about selling our oil off this continent, selling our energy products off to China," Prime Minister Stephen Harper said in response to American objections to Keystone. "I ran into several senior Americans, who all said, 'Don't worry, we'll get Keystone done. You can sell all of your oil to us.' I said, 'Yeah we'd love to but the problem is now we're on a different track.'"

Harper was referring to the Northern Gateway, a $5.5 billion pipeline with 525,000 bpd capacity that would cross westward to reach the Pacific Ocean and into the Chinese market. Sinopec is one of the leading investors of the project. Chinese companies have invested over $16 billion since 2010 in oil sands, and it's also a happy customer for all the planned increased output.

But like the Keystone XL pipeline, the western route faces staunch opposition over its environmental impact. Canada's targets "could be hindered by continuing debate about the environmental effects of oil sands exploitation," the IEA said in its 2011 WEO.

The Canadian government has announced it will improve monitoring of the environmental impact to address concerns. Oil sands are a mixture of sand, clay, water, and bitumen, most of which is a solid unconventional oil. The surface portion of the deposits, which currently accounts for most oil production, is literally mined. Tons of oil sands are trucked to complex refineries that first separate the bitumen and then upgrade it before it can be refined into consumer products. In the future, most of the remaining oil sands, those that are buried, will have to be extracted in situ by heating and diluting the bitumen to help pump it up in a more

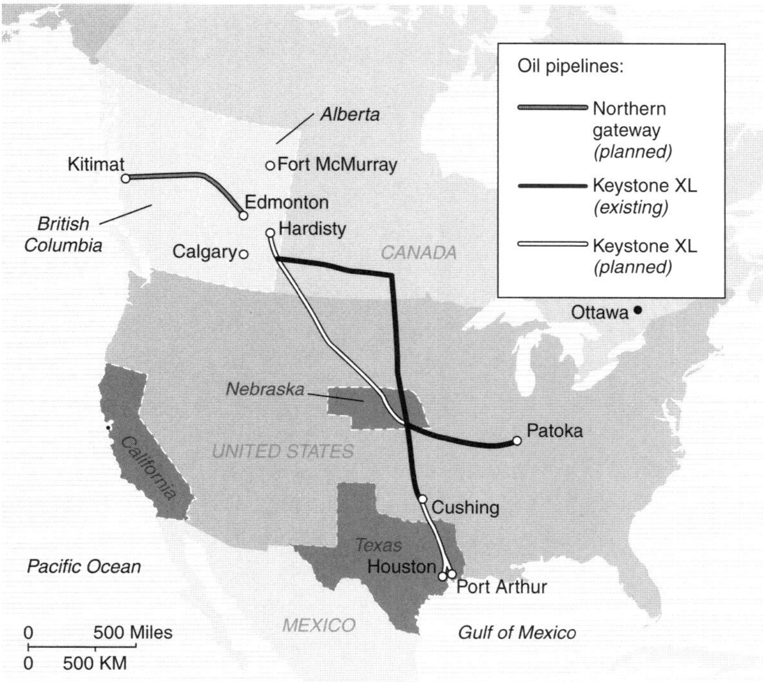

FIGURE 6.8 Existing and proposed pipeline connection to Canadian oil sands

Source: Canadian Association of Petroleum Producers

liquid form. But mining will still account for at least a third of oil sand extraction.

Surface mining of oil sands significantly alters the landscape and ecosystem because it literally implies opening holes in the ground several football fields big. It also produces toxic residues that are being accumulated in huge nearby pits and that still lack an economic means of disposal. There are over 65 square miles of tailings ponds, as they are known.[6] The process also requires large amounts of water and electricity, which means burning more natural gas and polluting more waters. Even with the cleaner in situ extraction, solvents that are added to liquefy the bitumen filter into the ground floor.

[6]Government of Alberta, October 2010.

Moreover, the EU Commission labeled products from Canada's oil sands as emitting 22 percent more carbon dioxide than conventional crude, in effect making them more expensive. Implementation of the rule continues to be delayed and will now require support from a majority of EU member states in a decision scheduled for the summer of 2012, although Canada is threatening to fight the measure internationally if it goes ahead. Europe is mirroring California, which already labeled fuels from oil sands as dirty, also shutting out its market.

These environmental, political, and economic limitations are compounded by the capital intensive nature of the investments, with longer return periods, all factors tending to delay production delivery. "Overall, the breakeven oil price for Canadian oil-sands projects is comparable to that of deepwater offshore conventional oil projects, but production, and therefore investment payback periods, is spread over a much longer time period," the IEA said in its 2010 WEO.

The IEA estimated in 2011 that most new oil-sands projects are profitable at prices "above $65 to $75 per barrel." But "investment payback is spread over a much longer time period. . . . As oil prices increase, some of the costs, notably of gas and services, will also rise, so the price threshold for profitability will also increase."

There are also bottlenecks that could delay oil sands, especially in human capital. There is already a deficit of qualified personnel, despite huge incentives. For each 1 million bpd of capacity developed, 20,000 qualified employees are required, according to the IEA. That is a recruiting nightmare. And the infrastructure requirements in Alberta to accommodate the newcomers will also take time to develop.

Still, Canada is looking to increase output from oil sands regardless of the challenges. At least one of the two pipelines will ultimately have to be built to open up new markets to Canadian crude, otherwise there would be no outlet for one of the world's biggest sources of additional oil. The question is how much politics delay these decisions.

An encouraging sign in March 2012 was Obama's decision to prioritize construction of the second leg of Keystone XL, that is, from Oklahoma to Texas. The bottom leg would increase profit margins for American and Canadian oil producers that have been offering significant discounts due to a supply glut in central United States. While the Obama administration's decision is still politically driven, the support is nonetheless realist in its ability to improve America's energy security. It might not support production from the oil sands now, but it will spur more output by making oil drilling more profitable in the United States and Canada. Early construction of the second leg will also minimize delays to the original Keystone XL pipeline plan, assuming the White House supports the first leg to Canada in the next administration.

Latin America, the pending policy

Canada won't alleviate oil prices much more this decade though, and any surprises are more likely to be on the downside. Latin America is the best positioned. It could theoretically more than double its oil production under different political circumstances, namely in Venezuela. Friendly investment policies in Colombia and Brazil will boost regional output as expected, but their potential is also constrained by available reserves. Instead of promoting investment in Latin America's energy sector with a coherent policy though, the United States has looked the other way.

Negroponte on improving US energy security:

> The single most important thing is North American integration of energy markets. But there is no silver bullet. We have to encourage as much projection as possible in this hemisphere. Whether in Brazil or wherever. But with Venezuela we're stuck with a populist Bolivarian political leader who is a little bit difficult to deal with. And yet he has not incompletely cut off ties to us.

Pumphrey concurs:

> In many ways the defining policy concern has been how to get a more diversified global market place. So how do you accelerate production increases in places like Brazil, like Africa? You have a demand policy to reduce demand for oil, but what we need to do is find a way to increase the number of sources so no individual region or country can become so dominant that you are basically held hostage to the dominance of a particular region or country.

Would it make a difference if Chávez can't hold on to power? Not necessarily in terms of oil production or global prices. The United States,

Chávez, and all of his potential successors share the goal of vastly increasing oil production in Venezuela. Chávez in practice is moving toward the center. It's an economic imperative, not a political choice. We will analyze why in the next chapters. But even if Chávez dies, the United States would still need to pursue a more productive relationship with the continent.

Just how much could Latin America, and Venezuela in particular, contribute to US energy security?

7 LATIN AMERICA: THE NEW FRONTIER

Latin America has the world's second largest crude oil reserves and most of them remain untapped. America's energy security is directly related to Latin America's ability to pump more oil, considering market prices, proximity, and regional influence, at least on paper.

A little more than half of the world's oil is in the Middle East. Latin America, from Mexico to Argentina, comes in second with around 18 percent, led by Venezuela, Brazil, and Mexico in that order. Cuba is not included, but its Gulf of Mexico fields—that could potentially hold about the same amount of oil as the United States—are not a game changer.

However, Latin America's share is expected to rise significantly as Brazil continues to certify its recently found subsalt fields, with indications that reserves could be exponentially more than currently reported. Venezuela also certified an additional 85 billion barrels in reserves early in 2011 which made its reserves officially the biggest in the world at nearly 297 billion barrels.

The American continent conservatively holds at least 30 percent of all known world reserves, including most of the world's largely undeveloped unconventional oil. The IEA is already very bullish about the continent, and in 2011 it forecast slightly less than half of the global incremental oil supply by 2020 will come from Canada, Brazil, and the United Sates. Venezuela is not expected to add much, assuming a pessimistic scenario in a Chávez-led regime.

American energy security in the short term will depend, as discussed in previous chapters, on bringing more oil output online on top of current expectations this decade as the economy crawls out of economic turmoil. And the only countries that could realistically add significantly more output—on top of current expectations—are Saudi Arabia and Venezuela.

Saudi Arabia has little motivation to unilaterally invest billions more in increasing its output capacity because falling oil prices would threaten its own economy. Riyadh has also rebuffed Western pressure in the past to invest more in production and to raise output. Policymakers have long acknowledged this reality and energy security has thus been about increasing output outside the Middle East.

In a business as usual scenario, depleted fields in Mexico and Venezuela will offset output gains mostly from Brazil and Colombia. In fact, the EIA expects the region's oil production to remain flat by 2020.[1] The IEA is a bit more optimistic, but regional gains are still capped at less than 2 million bpd by 2020. Meanwhile, demand increases could in fact outpace gains under current assumptions.

But what if Venezuela successfully implemented policies to boost production from depleting fields and the largely undeveloped Orinoco Belt? It could recover as much as 1 million bpd lost to inefficiency, lack of investment, mismanagement, and underskilled staff since Chávez took over. In addition, just by meeting already-signed development contracts, Venezuela could add at least another 1.5 million bpd this decade. Venezuela could conservatively add at least 2 million bpd on top of what it's already expected to contribute, although Chávez says he expects to double output of 3 million bpd to 6 million bpd by 2020.

The point is, global oil production gains this decade could conceivably be a third higher, that is, closer to 12 million bpd if Venezuela and the rest of Latin America beat pessimistic expectations. It's technically possible, but whether the politics are in place is a different question, and ultimately the most important one.

Oil south of the border

Mexico

Mexico's proven reserves are big at 10.4 billion barrels, but they are not huge. The United States has twice as much, and Canada and Venezuela hold exponentially more. But it's America's neighbor and any additional

---

[1]EIA 2011 Outlook: South America OPEC loses 500,000 bpd in conventional production, and additions from unconventional oil in Venezuela are "considerably uncertain"; Mexico and Chile lose 900,000 bpd, while Brazil and other countries add 1.5 million bpd.

production would directly benefit the United States, its natural export market. Total liquids production, including crude, NGLs, and refinery gains in 2010 was about 3 million bpd, a smaller decrease compared to the gradual production losses over the decade. Mexican production peaked in 2004 at around 3.8 million bpd of liquids, and output from its mature fields continues to decline. The Mexican government managed to stabilize plummeting production and has announced plans to recoup some of its losses for the remainder of the decade, although few believe that it is realistic. By 2020 the IEA expects output to fall to 2.5 million bpd.

Mexico's exports have been gradually falling on their way to 1 million bpd, but many in the industry believe it could become a net importer if it doesn't start adding production capacity soon. That will be impossible without major reforms. PEMEX, the state oil company, has a monopoly on oil production, and underinvestment, outdated technologies, and lack of exploration have severely limited Mexico's ability to boost production. In addition, Mexico's refining capacity is insufficient and its fuel demand is rising quickly, meaning an increasing portion of its crude exports are being reimported as refined product. Natural gas is also imported in northern Mexico from the United States.

Mexico has itself to blame. The country's constitution bars direct foreign investment, thus depriving the country of fresh capital and technology to boost production and reserves, which indeed most believe will increase once its deep waters in the Gulf of Mexico are adequately explored. Even Venezuela and Cuba offer a more investor-friendly energy sector, which says it all. The current government in 2008 managed to overcome legislative opposition to approve minor tweaks to its laws. These allow PEMEX to subcontract foreign operators for a fee, but not for a share of the production, thus limiting the reform's impact.

Politics has held back Mexico from realizing its oil production potential. Presidential elections are scheduled for July 2012, but none of the top contenders endorses the kind of reform that is required to change what appears to be an irreversible downward trend.

Negroponte on Mexico's regulatory problems:

> Mexico will soon be a net importer of oil. They will have to decide whether they really want to continue to exclude foreign oil companies from engaging in risk based investments in their oil sector. They will have to deal with that. There's nothing you or I can say to make

them do it. It's going to be an economic decision they will be made when they are faced with reality of traumatic decline in production. I think within the decade, by 2020 I'm sure, they will have to address the issue of whether or not exiting petroleum legislation satisfies their requirements.

In 2011, the United States still expected its total liquid imports from Mexico to fall just slightly through 2020.

Cuba

Cuba believes there are 20 billion barrels of recoverable crude oil in deep offshore fields of the Gulf of Mexico, adjacent to Florida. That would be equivalent to America's total reserves. The US Geological Survey (USGS) estimates there are more like 5.5 billion barrels, including NGLs, which would nonetheless be a significant find adjacent to the United States and its Gulf of Mexico refining capacity. Reserves that size are small compared to Venezuela's or Canada's, but Norway, with around 6 billion barrels in mostly offshore fields, produces 2.2 million bpd, almost as much as Nigeria, Africa's top producer.

Oil production, if a major discovery is made, could come online by the middle of this decade after exploration started in January 2012. Spain's Repsol was the first of several companies to drill exploratory wells in deep Cuban waters in the Gulf of Mexico, just a few dozen miles from Florida. Waiting in line to eventually use the rig are Norway's StatoilHydro, India's Oil and Natural Gas Corporation (ONGC), Russia's Gazprom, Malaysia's Petronas, and Angola's Sonangol. China National Petroleum Corporation (CNPC) and PDVSA are also heavily involved in Cuba's oil sector, from exploration to expanding refining capacity. And if any big discovery is made, the top global energy companies—except for Americans—are sure to seek their piece of the oil bonanza bordering Texas and Louisiana.

Companies will use the Scarabeo 9, a brand new, top of the line, Italian-owned and Chinese-made semisubmersible rig custom-made for Cuban exploration and designed to operate at more than twice the depth it will operate in. Any new output would contribute directly and immediately to American energy security, even if the embargo is still in place, because Cuba's demand is small and most oil would go to international markets.

There is nothing stopping the exploratory drilling other than that it's solely Cuba's decision. No law is being broken as the rig was tailor-made to make sure that less than 10 percent of the parts were American. There are no signs either that the United States intends to relax the decades old embargo anytime soon, which means that American companies will miss out on an oil boom on its border, at least initially. But if that wasn't enough, political pressure is mounting to try to stop Cuba from exploring, a self-defeating proposal that would damage America's energy security.

The small and disproportionally powerful anti-Castro Cuban American lobby in Florida is pressuring Congress to punish companies that invest in Cuba's oil sector. The argument is that the Castro regime will be propped up if oil is found and that the United States should impede anything that might make it more powerful.

Forget the fact that being communist or antidemocratic does not deter American energy interests elsewhere. The United States already imports almost 10 percent of its oil from Cuba's closest ally Venezuela. It makes no sense to thwart Cuban efforts to increase oil output perhaps in as little as three years, especially considering that oil prices will remain stubbornly high because global supply is constrained.

New production bordering the United States should be welcome news, especially if American oil companies are eventually allowed to take a stake and if environmental safeguards against spills are improved in the process. Indeed, the Obama administration, on the recommendation of America's oil industry and experts, appears set to work with Cuba on potential oil spill containment policies.

It's in the interest of the majority of Floridians, indeed a vast majority of Americans, to do so. American politics should not impede Cuba from getting access to the best spill containment technology and know-how, which not surprisingly is American, if nothing else because any oil spill would hurt the Florida coasts more than the Cuban ones.

Also, any unilateral punitive measure against investment in Cuba's oil sector could risk reciprocal policies that undermine American companies' interests in other countries. Worse yet, it could all but kill any hope of American companies getting a piece of what could be an oil bonanza.

Instead, Congress could seize this opportunity to relax the outdated embargo on the communist island, at least to allow US energy companies to partake in Cuban exploration and production. Phasing out the embargo would have bipartisan support, seconded by both environmental groups and oil companies, if the White House showed more leadership.

Cuba's oil future is uncertain of course until exploration is completed, and given the embargo, any serious offshore production could be delayed as it gets caught up in American domestic politics.

Colombia

Colombia doesn't have much oil, barring an overlooked find. It has 1.9 billion barrels of proved reserves, around 0.5 percent of Latin America's total reserves. Still, a significant upgrade is unlikely and Colombia at best might replace Argentina, with 2.5 billion barrels, as the fifth biggest proven reserve holder in Latin America, behind Venezuela, Brazil, Mexico, and Ecuador.

The importance of Colombia to American energy security is twofold. Geopolitically, it remains Washington's closest ally in South America, a particularly important role considering it neighbors Venezuela and Panama. More importantly though, Colombia is the fourth biggest oil producer in the region behind Mexico, Venezuela, and Brazil. It's also the third biggest source of American oil imports from Latin America, behind Mexico and Venezuela. And considering current trends, it could come to replace Venezuela as the second source.

Its production has soared year on year, leading the world in 2010 in terms of percentage gains as a result of improved security and huge foreign investment. In 2011, its output of nearly 1 million bpd was a historic record, its previous peak dating back to 1999 at 840,000 bpd. The government has set a goal of producing 1.4 million bpd by the middle of the decade.

Colombia's role in oil markets is also increasingly important thanks to its export capacity. Internal demand has remained flat below 250,000 bpd for the last decade, and even with expected demand growth as the economy roars, oil left over for export could rival Mexico's 1 million bpd this decade.

The bottleneck could come in pipeline capacity, which at 700,000 bpd is insufficient. By 2014, Colombia wants to double its capacity to 1.4 million bpd. China is taking a leading role in a proposed pipeline to connect Venezuelan oil deposits to Colombia's Pacific coast. The country also has an aggressive investment plan to boost refining capacity for its mostly heavy oil production. So far, much of the production is sent to the United States for processing, a portion of which is reimported.

Ecopetrol is the publically traded national oil company, 88 percent of which is state owned, although another 8 percent divestment is expected. More than 70 international companies operate in Colombia, attracted by "the establishment of a lower, sliding-scale royalty rate on oil projects; and longer exploration licenses," according to the EIA. However, major oil companies have yet to take up stakes in Colombia, partly because new production is mostly the result of enhanced recovery in declining fields and because no major new discovery has been made.

Violence remains a factor. Labor strikes in 2011 highlighted the risks involved in Colombia's oil sector, but overall, stability in the country appears set to continue improving in coming years.

Ecuador

Aside from Venezuela, Ecuador is the only other OPEC member in America. It rejoined the cartel in 2007 as its smallest producer following a 15-year absence. Reserves of 6.5 billion barrels are the fourth biggest in Latin America, but only equivalent to about 2.5 percent of the region's total.

Its production probably peaked in 2010 at 500,000 bpd, according to the IEA, as foreign investment in the oil sector has fallen following a wave of Chávez-inspired nationalization that has been compounded by regulatory and legal uncertainty, and OPEC quotas. More investment could reverse declines, but it's unlikely the country could contribute much more.

Internal demand has been rising rapidly over the past few years as half the power generation relies on oil. Around 40 percent of the production is consumed internally, a disproportionate amount considering the country's economy and population. Exports are increasingly under pressure as authorities try to attract investment and expertise.

In 2010, the government replaced the existing, more profitable production sharing agreements with service contracts. Windfall taxes were raised. Several big companies refused the terms and preferred to go. The assets of Occidental, a US company, were seized in a dispute with the government. However, the new operators lack the cash, and often the expertise, and production in some fields has fallen more than 50 percent. But there is little economic incentive and even Chinese companies are balking at the terms.

As a result of the government's decision to control the oil sector, forecasters expect production to continue to fall throughout the decade without major reforms. But Ecuador's production and exports are ultimately so small that they can easily be offset.

Brazil

Brazil last decade was arguably the biggest game changer when it comes to oil. Enormous financial and political capital was invested to radically increase oil output. It paid off handsomely with a 30 percent boost to production in the last decade alone, making Brazil a net exporter from 2009. Some of the biggest global oil discoveries in decades were made in the Florida-sized Brazilian presalt, or subsalt, layers deep in the Atlantic Ocean. The fields are so big, complex to explore, and new to companies that their size remains uncertain, with some estimates suggesting a recoverable reserves range between 100 and 150 billion barrels.[2] The government says Brazil will eventually certify more than 50 billion barrels of recoverable oil. So far, around 13 billion barrels have been certified as proven.

Brazil also has the financial and technological resources to develop the presalt deposits, plus a very attractive sector for international oil companies, many of which have long signed agreements. Repsol, British Gas, Galp, BP, Anadarko, and Sinopec have major interests in Brazil. Indeed, the bottleneck to access resources quickly is in the mammoth infrastructural requirements, starting with rigs.

The main player is Petrobras, the country's state-owned publically traded oil company. It is planning to invest $224 billion in the next five years to develop subsalt fields along with its international partners. Brazilian's output will more than double to 4.4 million bpd by 2020, according to the IEA. By law, Petrobras will keep at least a 30 percent stake in all future presalt fields.

Some estimates suggest Brazil could achieve 5 million bpd of crude production[3] by 2020, a quarter of which would come from presalt. The EIA expects a more conservative 3.3 million bpd.

American majors don't have a big presence in Brazil, although American oil equipment companies and subcontractors are playing an

---

[2]Hernani Chaves, Rio de Janeiro State University study, Center of Global Energy Studies (CGES) from London. Numerous press reports citing government analysis.

[3]CGES, March 2010.

ever increasing role. That could change when Brazil holds its eleventh oil and gas exploration bidding round for 174 onshore and offshore blocks in 2012, although it's not expected to include new subsalt fields until the government agrees on a development plan.

Obama has praised Brazil several times for its energy strategy, and has said the United States will happily buy all of Brazil's exports. That's more political rhetoric, of course. In reality, the United States is not budgeting for any serious additional imports from Brazil, according to the EIA 2011 outlook. In fact, most additional Brazilian exports are already heading toward China.

A further setback to American interests in Brazil were two accidental offshore deepwater oil spills in November 2011 and March 2012 in the Chevron-operated Frade appraisal well, 230 miles off the coast of Rio de Janeiro in 4,000 ft deep water. Chevron, which admitted responsibility, is facing criminal charges and a civil suit for over $22 billion, on top of millions in fines. The exact consequences will likely not be known for years, but the accident will probably add to expected delays from the country and American company concerns about regulatory liabilities.

The Frade spills were a fraction of BP's Macondo Gulf of Mexico accident and exploration was halted. The accident exposed a long-held fear in the industry: Brazil's oil output expectations this decade could be overblown and its ability to contribute much to global oil markets is uncertain.

"The spill shows that nobody was properly prepared, not Chevron, not the government, not the regulators," Reuters quoted Ildo Sauer, Petrobras's former top natural gas executive and a onetime energy adviser to Brazil's President Dilma Rousseff, as saying. "We need to totally rethink our safety paradigm. The whole over-optimistic, patriotic idea that offshore oil will solve all our problems is getting ahead of ourselves."

Regardless, the production potential is huge, no doubt, which explains why everyone wants in, and why Brazil is trying to keep much of it. Domestic self-sufficiency is welcome news for a tight market.

Some worry the government might be tempted to increasingly interfere, although there is little sign that investors are worried or losing interest. Infrastructural delays and lack of personnel, rigs, and other equipment could also delay development, increased internal demand could limit export capacity.

But the biggest challenges to meeting the target will be technical and regulatory. The technology for such complex and deep drilling is only a decade old, and while its development is a top priority and cash is not a problem, it's a new frontier in oil drilling that will require time to perfect.

Consider the following: most of Brazil's reserves lie at least 125 miles from the coast, more than 4 miles deep, under a layer of corrosive salt more than a mile thick. Aside from the logistical issues, submarine energy grids to bring the oil to the surface will have to be built and maintained at extreme depths and harsh conditions, including giant waves. The pressure of the subsalt oil fields is twice as high as regular conventional fields, which means equipment and steel able to withstand the corrosion and physical extremes of these deposits will have to be developed and built in huge numbers. Robots, submarines, and an array of new technology will be required, not to mention a vast number of rigs.

Brazil needs the crude and world markets want as much as it can spare. The first shipment of presalt oil was made in May 2011. What is in doubt is the ability of Petrobras and other foreign companies to develop the fields as fast as Brazil would like. Development costs of presalt are expected to range between $35 and $60 a barrel, with the high end competing with Canadian oil sand deposits.

In that sense, Canada and Brazil are in a similar position. Both countries have to overcome more technical than political challenges, and all efforts

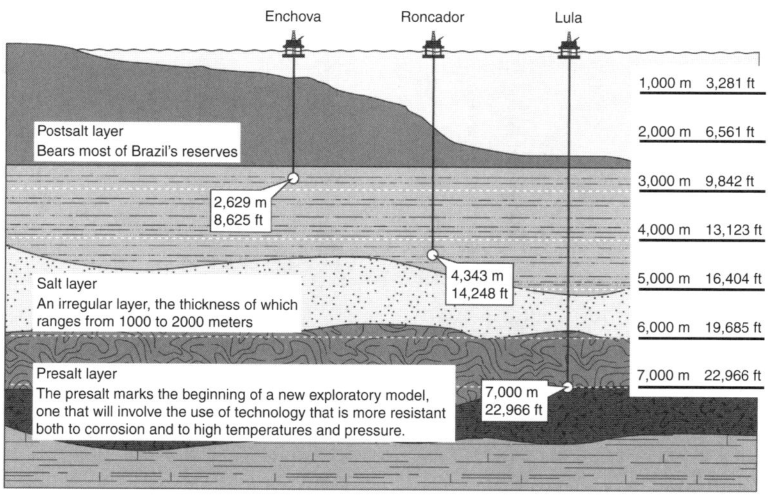

FIGURE 7.1 Geological challenges to Brazil's oil rich subsalt layer
Source: Petrobras

are in place to maximize production as fast as possible. In both cases, there are already huge expectations. But the bulk of oil won't be available until after this decade, too late to significantly alter the supply and demand balance to benefit American energy security now.

Venezuela, a silver bullet?

Venezuela holds around 46 percent of the continent's reserves, three-quarters of which are in the Orinoco Belt.

Luis E. Giusti, the last PDVSA CEO before Chávez's ascent, calculates Venezuela could hold about 25 billion barrels of light and medium proven reserves, another 70 billion barrels of conventional heavy, and perhaps another 50 billion from the Orinoco Belt. "For business it might be more like 70 or 80 billion barrels," he said in Bogota.

Venezuela is one of the world's biggest oil producers and exporters; its infrastructure is well developed, even if ill-maintained, its refining capacity is vast, and its interests and assets both in production and refining are spread throughout the world.

PDVSA, the state company, is not just one of the world's biggest energy companies, but a state within a state. It's involved in everything from fighting poverty to balancing the national budget. Indeed, the most significant challenge to Chávez's rule has come from PDVSA, not the political opposition or foreign governments.

But unquestionably Venezuela's oil wealth has been mismanaged by the Chávez regime. Venezuelan crude production plummeted after a two-month strike in 2002 that brought output to a standstill and it hasn't been able to recover.

Almost four years after Chávez took office in 1999, and only months after a failed military coup in April 2002, Venezuela's anti-Chávez opposition rallied support for a nationwide strike that included PDVSA and thus involved shutting in output. The stoppage started in December and lasted for around two months. Venezuela had to import crude not only for local supply but to meet export obligations. The entire export fleet was grounded, refineries and production stopped. The country lost \$13.3 billion as a result of the strike. The GDP for the first trimester of 2002 fell 27 percent.

Once the government reasserted control, it fired 18,000 employees, or 40 percent of the company's workforce. PDVSA management was fired. The leaders of the strike were issued arrest warrants and many fled. Chávez gutted PDVSA and reformed legislation to increase control over the company, including how it spent its money. Between 2004 and 2010, PDVSA contributed over $61 billion to social development, which came on top of government funds. Much of it went to finance Chávez's antipoverty program called Bolivarian Missions. In fact, PDVSA is also involved in dairy production and farming.

Chávez increased control of PDVSA also brought uncertainty to published production and export figures. He decided to stop filing numbers to the US Securities and Exchange Commission in 2005. As a result, bellwethers like *O&G Journal* and BP, and by extension the EIA and IEA, showed consistently lower output from Venezuela than the country claimed. The EIA says the difference is "largely due to measurement methodology." In 2009, to counter those agencies, Venezuela began publishing independently audited numbers. But in 2011 it announced it would no longer use the British firm it had hired.

The inconsistencies appear to be narrowing though. The IEA in 2011, in a special section of its Midterm Energy Outlook published in June decided to change how it collected information from Venezuela and to revise its numbers for the last decade. "With the divergence between official and third-party indicators," the IEA adjusted new parameters since the 2002 strike.[4] The result was Venezuela's output figures increased 500,000 bpd in 2009, and 300,000 bpd as an annual average between 2005 and 2011. The IEA and other bellwethers consistently report lower output than those reported by Venezuela, but at least there is no longer nearly 1 million bpd difference.

The United States and developed countries have long used plummeting oil production to expose Chávez as a bad leader. While the revisions don't change PDVSA's disappointing track record under Chávez's, they do reinforce suspicions in Chávez's quarter that the numbers are being politically manipulated.

---

[4]IEA Midterm Energy Outlook 2011: "Persistent concerns have surrounded the accuracy of production and export data emanating from official sources, following a debilitating strike at Venezuela's state oil company in 2002. With wide divergence between official and third-party indicators, we have undertaken a comprehensive review of data sources and decided to switch from a monthly estimate based on a proxy for wellhead production, to an assessment that combines observed international imports of Venezuelan crude with domestic refinery runs—a proxy for the broader measure of crude oil supply."

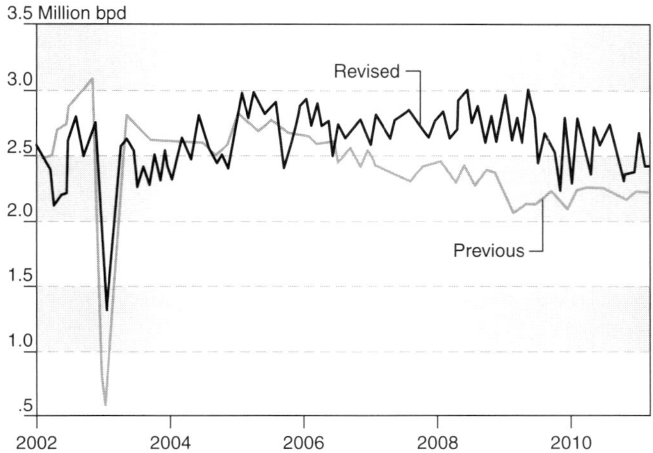

FIGURE 7.2 Venezuela oil production
Source: IEA Midterm Energy Outlook 2011

New production has come online from the Orinoco, masking the losses, but mismanagement is no doubt weighing increasingly on Venezuela. It's just not flowing as fast as some thought, and higher oil prices have further blunted the effects.

Venezuela's independently audited numbers show oil and products production of an average 2.8 million bpd in 2010, the lowest since the strike of 2002, of which it exported 2.3 million bpd. Total crude production is around 2 million bpd and the rest is products, mostly from NGLs and upgraded heavy crude from the Orinoco.

The *O&G Journal* put production at 2.4 million bpd for 2010 and exports for 2009 at 1.75 million bpd. The EIA estimates 2010 exports were even lower at 1.6 million bpd. BP and IEA separately estimated a production of 2.5 million bpd in 2010 and exports of 1.7 million bpd.

Venezuela has offered several explanations, including differences over how to account for unconventional production from the Orinoco. Some have also cited that the difference could lie in how much oil is exported at preferential prices to the Caribbean, especially Cuba.

In January 2011, Barclays Bank decided to do its own assessment and cross-checked different available numbers, including the value of imports from countries importing Venezuelan crude and destination reported by PDVSA. It estimated total oil exports from Venezuela of 2.4 million bpd in 2010 and 2.5 million bpd in 2009, which coincides with government audited numbers. "Our estimates show an average difference of just 6.2% since 2007 with respect to the exports reported by PDVSA," said the Barclays report.[5]

For 2011, PDVSA reported 3 million bpd and a target of 3.5 million bpd for 2012.

Most of Venezuela's oil is heavy and sour (or viscous and sulfur-rich), which is more expensive to process and to ship. It has to be diluted and mixed to pump around. And it's often blended with better crudes to improve its overall quality. Few refineries are geared for this type of crude and a big portion of them belong to PDVSA, with around 2.8 bpd of refining capacity spread throughout the world, mostly split between Venezuela and the Unites States.

US-based CITGO, which is wholly owned by PDVSA, has a refining capacity of 750,000 bpd, or around 4 percent of US demand. PDVSA also co-owns Hovensa refinery, with a 350,000 bpd capacity, and another 192,000 bpd from the Chalmette refinery.

Most crude comes from very mature wells around the Maracaibo Gulf, some with depletion rates—or annual production losses—of more than 25 percent. Reversing these declines requires huge investments of around $3 billion annually, according to the EIA. But Venezuela has been unable to sustain or even slow down production declines from many of its wells.

The main reason is underinvestment. PDVSA's revenue has been used as a piggybank to finance Chávez's revolution. It's also partly because Venezuela lacks enough natural gas production and shipping infrastructure to enhance recovery rates. It's also short on physical and human resources.

But all these issues can be addressed with sufficient cash and top of the line technology and expertise. That is why Venezuela is perhaps the best option to improve US energy security.

Total oil and product output from Venezuela peaked at nearly 3.5 million bpd in 1998, precisely before Chávez took office in January 1999. By 2011, production has fallen anywhere between 700,000 bpd and 1 million bpd, depending on methodology discrepancies, while adding at least 700,000 bpd of new production capacity, mostly in the Orinoco Belt. Those initial output gains have come mostly from easily accessible oil. The rest will be more challenging and requires new techniques.

---

[5]Barclays Capital | Venezuela: Venezuelan Oil Exports—Myths, Doubts and Realities, published January 31, 2011.

"The Orinoco could compensate for slower growth in Canadian oil-sands production, if Canadian projects were delayed by environmental concerns, provided Venezuela was more ready to accept international capital," the IEA said in its 2010 outlook.

In 2011, PDVSA invested $15.1 billion, 31 percent more than the year before. The company's debt also increased 40 percent to $34.8 billion. PDVSA's net worth is $75 billion, with booked assets of more than $162 billion, according to the country's Energy Minister Rafael Ramírez.

Looking forward, Venezuela has agreements in place, at least on paper, to add nearly 2.4 million bpd more of extra heavy crude from the Orinoco Belt this decade after investing $80 billion, which would translate to around 1.9 million bpd of upgraded crude. The IEA in its 2010 Outlook said it didn't expect capacity additions in the Orinoco to exceed 2 million bpd by 2020. In 2011, though, the agency said it expects less than 1 million bpd of output gains from the Orinoco by 2020. The EIA is even more bearish and expects basically flat production out of Venezuela.

These projections seem excessively pessimistic, especially because political considerations, not technical ones, are driving them. The new contracts will start delivering output by the middle of 2012, with drilling having already started in several blocks. Chevron says it will start producing in 2012. Other, like Italy's ENI, and Spanish Repsol have similar targets. Venezuela plans to invest $5 billion in 2012 in the Orinoco Belt to add 500,000 bpd. Most production will eventually come in association with national oil companies, mostly from China.

A great deal depends on Chinese companies, which have committed some $40 billion this decade to developing Venezuela's oil bounty through four different projects that involve oil production of about 800,000 bpd, and the construction of a refinery geared for Venezuela's heavy oil with a capacity of 200,000 bpd. Of course, the devil is in the details.

So far, Chinese companies have invested less than $1 billion, according to Derek Scissors, a research fellow in the Heritage Foundation who prepares the China Global Investment Tracker, one of the most comprehensive datasets on Chinese prowess abroad.

> Chinese find dealing with Venezuela very difficult. The Chinese government has a priority of diversifying oil supplies away from the Middle East with those from Latin America. But companies told the

government to take on the financial risk in Venezuela because they don't think Chávez will be able to pull it off. If it works, then companies will come in with the big money.

Still, the relationship is vital for Chávez. Venezuela has received some $30 billion in the form of loans from China that are paid over the long term with oil supplies. Chinese banks have also underwritten PDVSA refinery investments in Cuba.

Like with Canada, the concern is partly technical, although Venezuela's Orinoco Belt production is not as complex or expensive as Canadian oil sands, especially in deeper, more plentiful deposits. Costs are expected to be around a third cheaper,[6] in part because Venezuela's extra heavy oil is deeper, and thus at higher temperatures and pressure. It's easier to extract and the recovery rates are higher. The current expected output though is sure to increase with new technologies, both in Canada and Venezuela, although not in the short term.

The biggest problem is political. The government decided to increase windfall taxes yet again, which could change the financial incentives for companies to invest. Chávez needs more capital to fund his project and thus set a windfall tax depending on the price of oil. The system will move from a set royalty structure to a variable rate of 80 percent when prices exceed $70 a barrel, and 95 percent when they exceed $100 a barrel.

The problem, all agree, is that Chávez is simply not willing to play by market rules. The United States privately talked to Venezuela in the first years of Chávez. The United States wanted to find out "whether there are levers to convince them that it's in everyone global interests, including theirs, to allow greater scope for foreign investments to come in," Pumphrey said.

> We were trying to have a dialogue saying their policies limiting the role of international oil companies in favor of PDVSA were not in the long terms interest of global economy or their economy. Chávez clearly chose a pathway to give the state greater control.

---

[6]IEA 2010 WEO: "Primary production with multilateral horizontal wells, which gives higher recovery rates than in Canada, due to lower oil viscosity, is significantly cheaper. So, overall, assuming a mix of primary and steam stimulation, new projects would be expected to cost on average about one-third less than Canadian oil-sands projects on a per-barrel basis."

Giusti also contrasts the potential with the Chávez reality.

> Venezuela could be really the big answer for US oil demand in the future. It would take a lot of investment and a lot of recovery, but you have the reserves. A well laid plan can bring a couple million more in five years. But it will all depend on what happens in Venezuela. Chávez took over with 3.5 million bpd, plus 1.2 million bpd already in implementation.

At the same time, Venezuela has been diversifying its exports markets recently, in part due to lower US demand because of the crisis, but also as part of its strategy to find other markets in Asia. PDVSA said it exported nearly 2.5 million bpd in 2011, nearly 43 percent of which went to the United States, 26 percent to Asia, 25 percent to Latin America, and another 6 percent to Europe.

Also, there are early signs Chávez might be budging and offering flexible incentives to compensate for more profit-draining policies.[7] For example, ENI negotiated an alternative to the traditional vertical model in joint ventures in order to benefit from tax cuts. New Orinoco projects are also exempt from some of the new provisions.

Another promising sign, perhaps, was Venezuela's decision to offer better-than-market prices to attract foreign investors to develop its natural gas. Venezuela imports gas from Colombia, despite its huge gas reserves, and badly needs more domestic production, not only to fuel future power generation plants, but to increase production out of rapidly depleting wells.

Luis Giusti on the turnaround:

> After 13 years of poor results from its oil policies, and clearly motivated by the need of increased revenues, the Venezuelan government is pursuing ways of increasing production by inviting large foreign companies to participate in new developments in Venezuelan territory.

Italy's ENI and Spain's Repsol signed a deal with PDVSA in December 2011 to develop the Perla field, with 15 trillion cu. ft of reserves, only

---

[7]Carlos Bellorín, an IHS analyst in an article published in *International Association of Energy Economics*: www.dundee.ac.uk/cepmlp/gateway/index.php?news=31407

after Chávez agreed to pay $3.69 per British Thermal Unit, a 50 percent premium on the market price of gas at the time. The contract is long term and gas prices are sure to rise, but it's significant because PDVSA aggressively sought to attract foreign cash and expertise when left without better alternatives.

Still pending are at least 16 claims against Venezuela in several arbitration bodies, including the International Chamber of Commerce and the World Bank's International Centre for Settlement of Investment Disputes (ICSID) with an aggregate value of $21 billion. The outcome of these claims could weigh heavily on investor appetite in Venezuela as most of these are related to expropriations, especially in the energy field. ExxonMobil was awarded $907 million of the $7 billion it claimed, which Chávez claimed as a victory.

China's buying spree in Latin America

As discussed in previous chapters, China needs to secure more oil—and most other commodities—than any country in the world, probably for at least two more decades. Its efforts to do so also date back at least three decades in its own territory and gradually broadening the search to its surroundings, then to Asia, on to Africa and the Middle East, and eventually to America. It's ultimately a natural and expected evolution of a surging economy, especially one that has attained double-digit growth for years.

China is far from being able to compete with US long-established cultural, economic, and political ties in Latin America, but the success of its cash-for-oil diplomacy, especially over the past couple of years, has dazzled policymakers and diplomats alike. "They have every right to do that. But I did not and do not believe we should cede that to them, that we need to be also competing for soft power influence," Secretary of State Hillary Clinton said in an interview with *TIME magazine* November 2011.

Broadly speaking, China's foray started at the beginning of the last decade. Its first priority was merely diplomatic, literally buying off poor, weak governments to cut ties with its breakaway island of

Taiwan in exchange for donated infrastructure projects, much like it had successfully done in Africa. Those ties evolved rapidly as China began a long-haul commitment to forge strong economic ties to the region.

Bilateral cooperation deals evolved into economic memorandums of understanding and then into corporate agreements, which over the years quietly matured into contracts. China secured long-term commodity supply deals, while slowly taking up positions in production of everything from copper and coal, to oil. The big jump came amid the financial crisis. The strategy didn't change, but the circumstances did to the benefit of China. Surging economies of Latin America faced prospects of economic growth, albeit at a slower rate than before the crisis hit in 2008. But credit dried up and investment all but disappeared. All those economic development projects were threatened and the region desperately needed to keep cash flowing, or else face a compounded recession and political instability.

Many Latin American countries, starting with the smaller economies, were suddenly a lot more attracted to China and its cash. The United States and Europe were preoccupied with their own mess and had little investment or credit to offer. Diplomatically, the best the region could get is wishful thinking, if not indifference. Cue China's (cash) invasion of Latin America.

China was applying its version of last century's Western colonialism, except it had no political conditions. Business and all business. No human rights vindications, no democratic aspirations, no political meddling. China was interested in long-term stability and economic relations, not in transforming local countries. The hand fits the glove. Latin America needed China, China needed Latin America.

Even if Western countries had paid attention, there was little they could do to even slow the grasp of China's nearly $3 trillion in reserves. There was lip service, but little to offer in return. It had no money and just more conditions. The ground was fertile for China to root itself in the region and it thrived.

In 2010, Chinese foreign direct investment in Latin America was $15 billion, twice what it had invested in the previous two decades. China is the third investor in Latin America with a 9 percent share, behind the United States with 17 percent and the Netherlands with

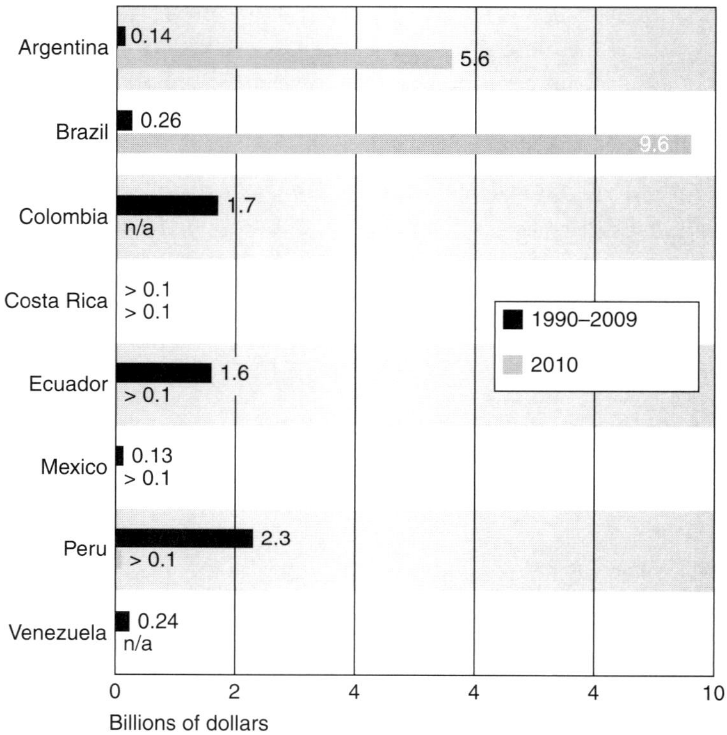

FIGURE 7.3 Chinese FDI in selected economies of Latin America
Source: United Nation's Economic Commission for Latin America and the Caribbean

13 percent.[8] By 2014, China will replace Europe as Latin America's second biggest trade partner after the United States, OECD's Secretary General Angel Gurría predicted.

Around 90 percent of its accumulated $45 billion invested so far, with billions more already agreed, has gone to the commodities sector, mainly oil. The region knew it had oil, but needed cash and expertise to develop, even if it still trailed Western know-how. Parallel to Chinese company investments, Beijing offered cash-for-oil swaps in Venezuela, Brazil, Ecuador, and several more countries. Billions in loans were

[8]ECLAC: www.cepal.org/prensa/noticias/comunicados/7/43307/hoja-informativa-China-es.pdf

signed, underwritten with long-term oil supplies. Brazil and Venezuela alone have received $40 billion combined. The money was critical and didn't affect debt levels as oil was the exchange currency. Venezuela on top benefited geopolitically from diversifying its markets.

"China's presence in Latin America has aroused great concern and anxiety. In fact, China's interest in Latin America is not to challenge the US dominance in the region, its 'back-yard,' but to promote South-South cooperation in the economic sphere. Naturally, as a region with a large market and an abundance of resources, Latin America is highly complementary economically with China," wrote Jiang Shixue, professor and vice president of Chinese Association of Latin American Studies.

Business ties with Venezuela are heavily weighed on by cash diplomacy, but most of China's inroads in the continent are through corporate, market-based channels. Chinese CNOOC bought a 50 percent stake in Argentina's Bridas Group, one of the biggest oil producers and refiners in the country. Sinopec bought assets owned by US-based Occidental, also in Argentina.

However, the most significant deals have been in Brazil. Early in 2011, Sinopec bought a 40 percent stake in the integrated Brazilian operations of Repsol, the Spanish oil major. Repsol has top of the line expertise and assets throughout the continent, from Alaska to Argentina, operating for over two decades in the region, and Sinopec is flush with cash. Months later, Sinopec also bought a 30 percent stake in the Brazilian unit of Portuguese oil company Galp. China's Sinochem agreed to buy a 40 percent stake in a Brazilian offshore oil field operated by StatoilHydro.

China's energy incursion into Latin America is not concerning from a market point of view. In fact, it's good news for the United States that somebody has been able to invest the billions of dollars needed to develop oil resources of Brazil and Venezuela. Without it, the amount of oil coming online in the next few years would be much smaller, and the supply and demand balance would be even tighter, meaning prices would no doubt be higher. China thwarted a deeper economic downturn in the region by investing in oil capacity while developed countries and Western companies were bunkered down.

As already analyzed though, there is also the political and geopolitical concern. How does China intend to manage the great degree of influence it has garnered in the region in the past few years?

Giusti sees a natural evolution of China in oil markets.

> If you are going to be a big oil consumer, you better be a big player as well. For China, it's different to be a simple buyer of oil than to be a producer of oil, to have international refining, to have penetration in the US market, because when you put that mosaic together you are going to be a completely different player and that's probably what China is looking at. There is no contradiction. The oil business is a global thing. Countries have policies, but in the end it is oil companies that do things. What I see is a playing field with an increased number of actors. I don't see a threat to the US.

Furthermore, China's influence is minute in the region, compared to the United States. The United States might be tied down now, but for decades it was basically the only player. It will be some time before China can exert more pressure in Latin America than the United States, especially considering Brazil's rise, which will thwart not just Chinese, but American aspirations.

Security and intelligence analysis firm Stratfor evaluated the trend with Brazil as an example:

> In the wake of the global financial crisis that began in 2009, Brazil experienced a significant boom in both exports and investment. Trade with China is the main driver of Brazil's export growth. China has become Brazil's largest export destination, primarily importing raw commodities—soybeans, iron ore and crude oil—and their byproducts. Brazil's exports to China in the first seven months of 2011 totaled $24.4 billion, or 17 percent of total exports (an increase from the same period last year, during which Brazil's exports to China totaled $16.7 billion and 15.7 percent of total exports). Brazil's relatively positive economic outlook and stable political system made it one of the most popular destinations for foreign portfolio and direct investment. Foreign direct investment alone amounted to $48.5 billion in 2010, an increase of more than 480 percent from 2003.
>
> The huge influx of foreign capital and large trade surpluses has driven up the value of the real. From mid-May 2010 to late July 2011, the real increased by 18.6 percent against the dollar and currently hovers at $1.65 per real. Though the strengthening of the real has a limited effect on producers of commodities (which are traded in dollars), it has directly and negatively affected the competitiveness of Brazil's

value-added manufacturing sector, particularly in combination with lowered global demand for manufactured goods.[9]

Hard choices ahead

Bottom line, US national and energy security this decade will largely depend on Latin America's ability to pump more oil, a lot more oil than it is already expected to contribute. But the ability of the American government to influence that, especially considering its long-held laissez-faire policy, is limited, as we'll show in the following chapters. Involvement of American companies in the region, which would no doubt have the ability to increase production, especially in Venezuela, is ultimately a decision in which policymakers at best can nudge, not push.

The United States can do little more than improve relations with the region, or at least not complicate them further with diplomatic policies like embargoes and sanctions. That is a political decision, not an economic or technical one.

A new approach is required. It will inevitably involve dealing with Chávez, who will likely extend his mandate, unless his health gets in the way. Efforts to alienate the anti-American president have backfired and now even Colombia, the most important American regional ally, is moving to seek its own stable relations despite Washington's concern.

Is Venezuela a silver bullet for oil markets, and indirectly to US energy security? The quick answer is that it easily could be, perhaps more so than any other country in the world. It's underproducing by as much as 1 million bpd, but it could rather quickly recover a big part of that with new technologies, money, and know-how. And if the Orinoco Belt is well managed, it could add at least 2 million bpd this decade. With a more inviting energy sector, one that attracts oil majors, current production could indeed double to achieve Chávez's goal.

The technological challenges are surmountable. This could theoretically happen. But it would imply recalibrating American strategic policy toward Latin America. What could the United States do to spur oil production in Venezuela and the rest of Latin America, without having to sacrifice American values like the defense of human rights, of democratic values, of freedom of the press?

---

[9]Stratfor: www.stratfor.com/analysis/20110908-brazil-responds-economic-slowdown

PART THREE

A NEW APPROACH

8 COLOMBIAN LESSONS FOR OBAMA

In May 2008, President Barack Obama assessed US relations with the Latin America[1] during the campaign trail. He spent part of his childhood in Indonesia and visited several countries in Africa, the Middle East, Eastern Europe, and Asia, mostly as a Senator, but he hadn't set foot in Latin America.

> Since the Bush Administration launched a misguided war in Iraq, its policy in the Americas has been negligent toward our friends, ineffective with our adversaries, disinterested in the challenges that matter in peoples' lives, and incapable of advancing our interests in the region.
>
> No wonder, then, that demagogues like Hugo Chávez have stepped into this vacuum. His predictable yet perilous mix of anti-American rhetoric, authoritarian government, and checkbook diplomacy offers the same false promise as the tried and failed ideologies of the past. But the United States is so alienated from the rest of the Americas that this stale vision has gone unchallenged, and has even made inroads from Bolivia to Nicaragua. And Chávez and his allies are not the only ones filling the vacuum. While the United States fails to address the changing realities in the Americas, others from Europe and Asia—notably China—have stepped up their own engagement. Iran has drawn closer to Venezuela.

Around the same time and echoing many Latin American leaders and academic publications, the Council on Foreign Relations task force report offered Obama a policy roadmap to relaunch relations with the region:[2]

> Achieving U.S. objectives and protecting U.S. interests in the Western Hemisphere requires an unsentimental and reality-based assessment

---

[1]Speech delivered to the Cuban American National Foundation in Miami, May 23, 2008.

[2]May 2008: www.cfr.org/mexico/us-latin-america-relations/p16279

of the complex and dynamic changes under way in Latin America and in U.S.-Latin America relations—and of the ways in which the United States can influence those changes for the better: Latin America's citizens rightly expect democracy to deliver more equality, social justice, and prosperity—not just formal representation.

And yet, aside from a few historical updates, both the assessment and policy roadmap could just as easily apply to the Obama administration after nearly four years in office, and to the next president, whether it's Obama or a Republican challenger. Relations south of the border are still circumscribed to drugs, immigration, and expanding FTAs. Policy debates are more about how to respond to a gang of bad-mouthing populist presidents, instead of how to reverse growing disenchantment in countries like Brazil, Mexico, Argentina, Chile, and Colombia. Latin America has matured into a critical long-term interest strategically, economically, and even militarily, while America remains limited by self-defeating, outdated priorities.

It doesn't need to be that way. Obama's detachment is optional. As we'll analyze in this chapter, he fumbled the Honduran diplomatic crisis to Chávez's benefit; Colombia U-turned on its foreign policy to seek closer ties with Venezuela; foreign powers are nesting in the region; Brazil consolidated its hegemonic power role; and above all US policies are having the unwanted effect of reinforcing Chávez politically while limiting his ability to increase oil production.

Brazil's outgoing President Lula da Silva perhaps best reflected Latin American frustration with the Obama administration just before his term expired in December 2010. Lula was speaking not only as Brazil's president, but as Latin America's most popular leader at the time, internally and regionally, and as the architect of the economic and political model for most of the region.

> US vision of Latin America changed little to nothing and that saddens me a lot because we are witnessing the most important democratization process in the world. I would like for US relations with Latin America to be different from what they are today.[3]

---

[3]Lula was widely quoted as saying during the traditional end of the year briefing with journalist December 27, 2010.

Obama's promise to "reestablish American leadership in the hemisphere" fell through. Foreign policy toward the region is just a tweaked version of Bush's legacy, which in turn was a botched extension of the mess that Bill Clinton left behind. For at least a couple of decades, this blind spot has quietly and gradually eaten away at American interests in this vital region, and this administration is making it worse precisely when more is at stake.

In the meantime, the spread of Bolivarianism has peaked, no thanks to Washington, but to Chávez's disastrous economic management. It's true Chávez reduced poverty, thanks to sky-high oil prices, but his economic policies are proving unsustainable, without mentioning the authoritarian tilt and political instability.

The recent ascent of former Chávez wannabe Ollanta Humala to the presidency of Peru illustrates this trend. Humala is also a nationalist, populist, and left-leaning leader, but he openly rejected the Venezuelan model, and embraced Brazil's, to avoid scaring away voters. New Colombian President Juan Manual Santos also said he wants to mirror Colombia's development after Brazil and Chile, and that Lula is his role model. Few are looking to the United States for leadership.

Former Colombian Vice President Santos on Venezuela's model:

> The problem for Chávez is that nobody pays attention. He is increasingly an irrelevant actor internationally. People don't believe him anymore. His promises didn't materialize. So the more the US fights Chávez the stronger he gets.

To be sure, while the United States and Chávez continue to offer diametrically opposed models for Latin American, most countries have concluded that neither offers a viable future, choosing instead a middle ground based on a political and economic balance between pro-market and statist forces. And even if Venezuela stands to lose more, in the short term this status quo does not bode well for American energy security or economy, and by extension, for its national security.

Polls only confirm a trend that was obvious to the naked eye: under Bush, the image Latin Americans had of the United States plummeted, then soared in the first year of Obama, and dropped again in 2009 and 2010.[4]

---

[4]Global Attitudes Project of the Pew Research Center

Latin Americans were hopeful that the "Yes, we can" message would help bury eight years of what most considered absent, often arrogant policies toward friends and foes alike.

It's not a question of ideology, as many in Washington oversimplify, but of engagement. As long as Latin America remains a second-tier priority, policies will likely remain trapped in a failed post-Cold War mentality, kidnapped by simplistic anticommunist ideas of good versus evil. Chávez is the exception, not the norm, and America's regional policy cannot remain hostage to what he does. The priorities should be set by the biggest economies: Brazil, Mexico, Argentina, and Colombia, not by the ideological fringe, Venezuela.

The United States of course remains the single most influential outside player in Latin American affairs. The Pew Research Center year after year shows Latin Americans have more favorable opinions of the United States than other regions. But they also suggest its influence is waning. The United States remains the biggest single investor, but its share has been falling. In 2000, 43 percent of all major FDIs in Latin America came from the United States. Between 2006 and 2009, that share fell to 25 percent, and in 2010 it fell again to 17 percent. It's true that OECD members in general have cut investments, but the United States has a lot more to lose, especially as other countries fill the Western void.

Indeed, FDI in Latin America over the past decade increased more than 50 percent by 2010 to $113 billion, while outflows reached a record $43 billion, reflecting the rise of intra-Latin American flows. This is despite the global crisis, in which Latin America "has been the most dynamic, with the greatest increase in both inward and outward FDI flows," according to the UN's Economic Commission for Latin America and the Caribbean (ECLAC).[5]

While intraregional share of FDI has gradually increased, it's still a discreet 10 percent of the total, but global trade, especially with China, has increased remarkably. Chinese FDI in Latin America in 2010 was the third biggest in the region at $15 billion, accounting for 9 percent of the total, and more than twice the combined investments between 1990 and 2009. Chinese FDI in the region decreased to $8 billion in 2011, while

---

[5]ECLAC, 2010 FDI report: www.eclac.org/publicaciones/xml/0/43290/2011-323-LIEI-WEB.pdf

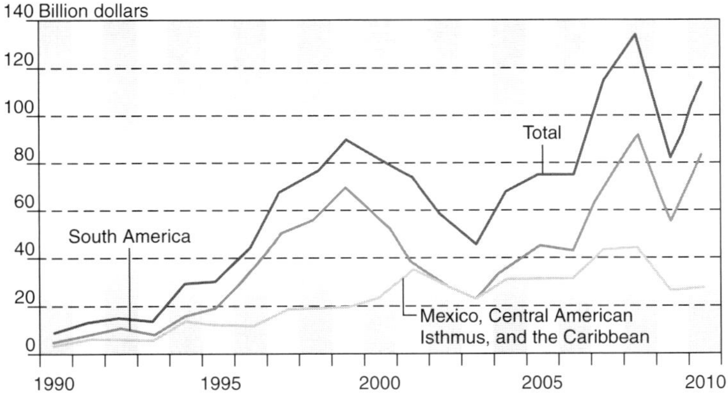

FIGURE 8.1 Latin America and the Caribbean: FDI inflows by subregion, 1990–2010

Source: ECLAC on the basis of official figures and estimates

total FDI in the region reached a record $153 billion. Venezuela received $5.3 billion totally, three times more than the previous year, 65 percent of which went to the oil sector, although divestments continue to outpace capital investments, according to ECLAC.

China is also Latin America's third biggest trading partner, after the United States and the European Union, but it will become the second by 2014, replacing Europe. China is an important market, accounting for 23 percent of Chile's exports, 15 percent of Peru's, and more importantly 13 percent of Brazil's. It will become the second biggest buyer of Latin American products after the United States in 2014. China exports mostly medium and high tech products, with Latin America sending raw materials, mostly crude oil, metals, and vegetable oils.

But instead of promoting investment in Latin America, especially in the energy sector in which many others are finding lucrative returns, the United States remains stuck debating how much of a threat Chávez is, an irrelevant issue considering what's really at stake.

There are two opposing versions, one of totalitarian, communist, unbalanced Iranian wannabe that supports terrorism, and the other of a thorn in the side, populist big mouth that is nothing more than a pain, far from posing any real threat to the United States. The reality

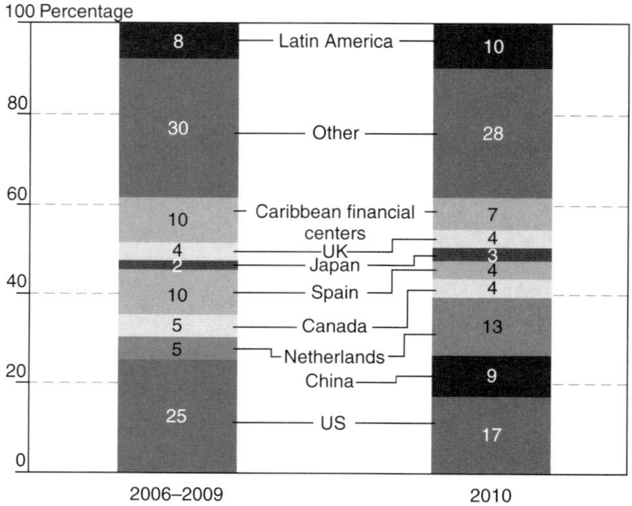

FIGURE 8.2 Latin America and the Caribbean: Origin of FDI inflows by subregion, 2006–2010 (the distribution of FDI by origin shown in this figure accounts for 80% of all FDI)

Source: United Nation's Economic Commission for Latin America and the Caribbean

is in the middle, and policies should be designed around that middle ground.

Chávez remains the most popular leader in Venezuela. Starting late in 2010, he has ideologically tried distancing himself from his more radical persona. It could be that he wants to show his milder side to the disenchanted middle classes ahead of the 2012 presidential elections for another six-year term. Perhaps he wants to attract more foreign capital to develop his economy, especially ahead of elections. Or perhaps his illness has softened him.

The more likely explanation is that Chávez is morphing to adapt to a new political reality in the region and the country. He has to be less radical if he wants to be reelected. He also needs better relations with his neighbors to bolster integration. And he needs to show his Bolivarian revolution is strong and stable to attract capital, especially now that Venezuela's future is uncertain because of his health. Even investors like China are increasingly worried about Chávez's political survival, and thus want a more amenable Chávez.

More importantly, Chávez's motives are irrelevant at this juncture. The United States should seize the moment, not further alienate itself for petty political gains ahead of elections in both countries.

Obama's blunders

Three months into his presidency, Obama visited Mexico in April 2009 as a stopover to attend the Summit of the Americas in Trinidad and Tobago. Regional leaders there gave Obama an earful of demands but their words fell on deaf ears. First, hopes that the United States would move to end the embargo on Cuba, a historically divisive issue with the rest of the hemisphere, withered. Obama was concentrating on domestic affairs like health care and handling the economic crisis, and the administration had little time or interest in rocking the boat, least of all with the Cuban issue.

Obama's roadmap for relations was just a rehash of Bush's policies, tweaked with a soft power approach. Relations would be based on equality rather than patronage. In reality though, only Mexico, which he visited again in August 2009, got much attention, and then again as a domestic issue in terms of shared economic, immigration, and security interests.

The first real test for Obama came with the military coup in Honduras that forced pro-Chávez President Manuel Zelaya into exile on June 2009. Obama said then:

> The region has made enormous progress over the last 20 years in establishing democratic traditions in Central America and Latin America. We don't want to go back to a dark past. The United States has not always stood as it should with some of these fledgling democracies, but over the last several years, I think both Republicans and Democrats in the United States have recognized that we always want to stand with democracy, even if the results don't always mean that the leaders of those countries are favorable towards the United States. And that is a tradition that we want to continue".[6]

Obama didn't recognize the new government and withheld aid. Honduras was suspended from the OAS, but the hemisphere split into three camps,

---

[6]Obama during a press conference June 29, 2009: www.whitehouse.gov/the-press-office/remarks-president-obama-and-president-uribe-colombia-joint-press-availability

with the United States placing itself at odds with the majority. All this was happening while Obama's choice for top diplomat for Latin America was blocked in the Senate, leaving a political vacuum that was sorely felt.

Washington-backed negotiations for a quick return to democratic rule failed, and the Obama administration wobbled with contradictory, lukewarm responses that satisfied few inside or outside the country. The United States, along with Colombia, Costa Rica, and other smaller countries, rushed to recognize a new president elected in November 2009 before Honduras had resolved the legitimacy of the new government or the standing of one that remained in exile.

A second camp, led by Chile and Mexico, delayed its recognition until August 2010 once the OAS confirmed Honduras was working toward reconciliation through a series of measures, including allowing the return of the ousted pro-Chávez leader Zelaya. The third and largest camp, including Argentina, Brazil, Venezuela, and five other countries, withheld recognition and blocked Honduras's return to the OAS until all criminal charges on the deposed leader were dropped in June 2011, clearing the way for his return and a national unity transition.

Chávez beyond a doubt entrenched himself in Honduras, using his oil in support of his ally, who had joined the Venezuelan-led ALBA alternative, triggering the military coup. But the majority of Latin America did not share America's conclusion that a radical shift toward Bolivarianism justified support for a military coup.

Former President Uribe of Colombia recalled the Honduran coup during the Bogota interview:

> One of the determinant causes of the Honduran crisis was a result of Venezuela's role when oil prices were high. Oil prices can bring serious risks to democracy in some countries and in others when they take advantage of those prices to remain popular or to push colonialism in countries that have no oil.

But the rest of Latin America had a different interpretation. It wasn't going soft on Chávez. It had just moved on after concluding that the *caudillo* was harmless as long as his failed policies exposed his unsustainable model. The region determined that Washington, on the other hand, was on the wrong side of history, and that its policies were propping Chávez up geopolitically and raising his profile unnecessarily precisely when he was most desperate and when he least expected it. Chávez came to

Honduras's defense and captured the impasse as his. Chávez made it about US imperialism and Washington was stuck defending a failed policy.

Obama's administration put its priorities at odds with all big countries, except Colombia. Stopping Chávez was more important than an unequivocal defense of democracy. And Latin America interpreted this as a serious setback, a reminder of Bush's stance on the Venezuelan coup in 2002, and of a not so distant past.

America's response though wasn't driven by a calculated interest, but by the domestic leadership vacuum. The confirmation of Arturo Valenzuela as Assistant Secretary of State for Western Hemisphere Affairs was delayed until November 2009 by Republicans who were at odds with Obama over the handling of the Honduran crisis. And even then, Valenzuela, a well respected diplomat and academic, simply lacked the political support to actually implement any meaningful policy reforms.

In March 2011, Joseph S. Tulchin, a visiting fellow at the David Rockefeller Center for Latin American Studies at Harvard University and one of the best-known Latin American scholars, analyzed Obama's policy toward the region and the challenges it faced.

> What was really significant about the Honduras episode is that it pushed the Obama administration further away from engagement with Latin America. The administration was, after all, completely swamped by two wars, the greatest economic disaster since the Great Depression and a variety of other concerns. In addition, the Secretary of State was showing herself to be extremely averse to risk taking. Even with the confirmation of Arturo Valenzuela as Assistant Secretary in November 2009, himself a cautious policymaker who would not push his boss to take risks in this complex situation, there appeared to be no reason to put Latin America any higher on the administration's agenda. Mexico, although different, was treated as if it were a domestic political problem, not a foreign policy issue, and was handled outside the State Department. Because of the economic crisis, trade policy was put on the back burner. The discussion of the strategic partnership with Brazil did not go very far because the drive by Brazil for a larger role in international affairs left observers in official Washington nonplused. Specifically, President Lula's joining with Turkey to offer a compromise deal with Iran over nuclear issues, precisely as the United States was pushing in the United Nations for further sanctions against Iran, so disturbed the administration

that Obama spoke during his visit to India of supporting India's pretentions to a seat on the United Nations Security Council, the very jewel of the crown Brazil hoped to wear.

Latin America fell off the radar screen throughout most of 2010. American regional policies wavered, and once again focused on Chávez and on getting Honduras back into the OAS. Meanwhile, Colombian and Venezuelan relations were at their worst, while China was making inroads in Brazil, Venezuela, and Argentina.

Around then, Juan Manuel Santos, Uribe's former defense minister and constant target of Chávez attacks, was elected president of Colombia, promising continuity with Uribe's administration. Taking office in August 2010, he immediately made it clear domestic policies would indeed consolidate Uribe's security and economic gains. But his foreign policy took a huge and unexpected U-turn.

President Santos unexpectedly decided to co-opt, rather than confront, Chávez. The two leaders quickly reactivated diplomatic relations and frozen bilateral trade. Ecuador too reestablished relations severed when Colombia bombed its territory to kill Reyes, the FARC leader. Santos prioritized ties with the rest of South America and called for more regional integration. His popularity grew and he became the new darling in South America. Without hurting US ties, Colombia set its priorities south, not north to the United States.

In fact, it was Santos's partnership with none other than Chávez that finally resolved the Honduran impasse, two years after the coup. Only a month after Obama's trip, the two brokered an agreement in April 2011 between the two rival Honduran sides and paved the way for the country's return to the OAS.

Santos is not warming to Bolivarianism. Nothing could be further from the truth. Colombia retains a right to center tilt and is still one of the most pro-market supporters in the continent. But Colombia's economy and counterinsurgency war have priority over United States ties that offered little in return, not even an FTA. We'll analyze the Santos approach further later in this chapter, but suffice to say the unexpected strategic shift has paid off handsomely, not just with Latin America, but with the United States. It's a pragmatic approach, one that delivers dividends, and in the process co-ops Chávez and undermines his more dangerous, radical version.

Santos's diplomatic turnaround also propelled the founding of the Union of South American Nations, or UNASUR for its Spanish acronym.

The nascent organization came to be in March 2011 and is envisioned as a precursor to a European Union-like integration. All South American countries have ratified the treaty and Mexico has asked to join as an associate member. There are already immigration agreements to forego tourist visas, and some countries are moving to relax labor mobility. The body has worked to promote intraregional alliances, from the Mercosur FTA, to building energy connection infrastructure. In addition, the countries formed a joint security and defense forum to share information and promote cooperation, although it is not a NATO equivalent. And the United States is not involved.

"During these times of change that the region is living, with the strength of a continent that awakens, we are heading to achieve our second and definitive independence," said Ecuadorian President Correa in reference to the United States in the inaugural ceremony of UNASUR. Of course, the results of such grandiose plans remain to be seen. But Latin American unity has already weakened the larger continental OAS body that includes the United States and Canada, while Mexico is splitting its bets.

Obama returned to Latin America, first to Brazil, then to Chile, and finally to El Salvador, in March 2011. He had traveled around the globe and visited every other continent, but had never set foot in South America. Obama's goal was to reinforce strong ties, but in reality he was playing catch-up. Not surprisingly, his trip accomplished little, except perhaps addressing domestic policy.

There were no hard feelings, but South America said it was doing all right by itself, with Brazil as regional hegemonic power and a wide spectrum to choose from between right and left. It's not that South America is now a bastion of stability and prosperity. All the problems of the past remain, but it feels comfortable enough to deal with Chávez and more pressing economic issues without American supervision.

Two months later, in May 2011, the US government finally decided to take on Chávez and probably in the worst possible way: toothless sanctions that barely scathe his regime, but consolidated him in power. Broader US interests in Latin America were clearly not the priority.

PDVSA, the oil mammoth and one of America's biggest oil suppliers, was included in the list of seven companies supporting Iran's energy sector in violation of US unilateral sanctions. PDVSA "delivered at least two cargoes of reformate to Iran between December 2010 and March 2011, worth approximately $50 million," the Department of

State[7] said in May. Reformate is blended with gasoline to improve its quality. The sanctions also bar PDVSA from access to US government contracts, financing through the Export-Import Bank, and export licenses. But the sanctions don't apply to PDVSA subsidiaries and don't "prohibit the export of crude oil to the United States."

Chávez could not hide his joy on Twitter: "Sanctions . . . imposed by the imperialist gringo government? Welcome Mr. Obama! The real impact of the new gringo aggression is potentiating the patriotic and nationalist moral of Venezuela." Chávez was right. Ahead of presidential elections in Venezuela, perhaps the best chance for a democratic transition of power, the Obama administration handed its biggest antagonist on the continent an ideological missile to rally support against big, bad Uncle Sam. "The revolutionary government calls on all the Venezuelan people, laborers and especially the oil workers, to stay alert and mobilized in defense of our PDVSA and the sacred sovereignty of the homeland," the Venezuelan foreign ministry replied.[8]

The $50 million worth of gasoline additive supply is a drop in the ocean for Iran, and Venezuela's ability to supply Iran with refined products is inconsequential. Venezuela is hardly able to meet local demand, while it's easier for Iran to get supplies from huge Asian refiners. In other words, the United States swallowed the bait.

In a written Q&A with Venezuelan newspaper *El Universal* in November 2011, Obama wrote:[9]

> The United States does not pretend to dictate its foreign affairs. I would argue, however, that the Venezuelan government's ties to Iran and Cuba have not served the interests of Venezuela or the Venezuelan people.
>
> Here in the Americas, we take Iranian activities, including in Venezuela, very seriously and we will continue to monitor them closely.
>
> The United States has already taken a number of significant and effective steps to indicate our concern to the Venezuelan government, including annual certification of Venezuela for not fully cooperating with anti-terrorism efforts each year since 2006. Most recently, we imposed sanctions on PDVSA for selling gasoline components to Iran.

---

[7]www.state.gov/r/pa/prs/ps/2011/05/164132.htm

[8]http://venezuela-us.org/2011/05/24/venezuela-rejects-u-s-sanctions-against-pdvsa

[9]www.eluniversal.com/nacional-y-politica/111219/venezuelan-govts-ties-to-iran-and-cuba-have-not-served-its-interests

In any case, the sanctions do not bode well for America's energy security, especially if they threaten future investments in Venezuela's energy sector, and if the decision was condemned by the rest of Latin America. The United States is only isolating itself further and undermining its energy security in its own hemisphere. Latin America wants the United States to lift an embargo on Cuba, not to pursue a new standoff with Venezuela.

Assuming just an honest, naïve mistake, it could be that the Obama administration concluded that the priority was sending a message to foreign companies over circumventing sanctions on Iran, and that the consequences on PDVSA or Venezuela's oil production were secondary. It could be that Chávez calculated the move to provoke the United States but force it to appear weak or hamstrung by economic considerations. Or it could be that it's a political calculation to shore up Cuban American votes in Florida, or to appease Republican demand for a harder line on Chávez.

But PDVSA could run into more trouble down the line if it becomes harder to find spare parts for its oil industry. It won't be eligible for US government loans, although Chávez never intended to request one. Regardless, the sanctions are unwise and add uncertainty to oil markets that are wondering whether the United States is moving to impose more crippling sanctions on Venezuela's oil industry, especially with the precedent set with the Iranian quagmire. Western governments have exposed their willingness to use oil as a weapon, and messing with the market only increases prices, not the other way around. It raises investment risks, including for American companies still heavily involved in Venezuela like Chevron. They don't dent Iran's refined product supply. They are not "a strong signal" to Venezuela, as the Department of State suggested. They strengthen Chávez ahead of elections. And they achieve little more than raising oil prices that threaten US economic recovery, while helping Russia, Iran, and Venezuela.

Further pressure could be catastrophic. Full sanctions would require replacing heavy oil supplies that much of the US gulf oil refining capacity is geared for, which Canadian crude could do, but not right away. In addition, oil market prices could soar as a result, an unwelcome scenario in a still fragile recovery.[10]

---

[10]Government Accountability Office, *Issues Related to Potential Reductions in Venezuelan Oil Production*, June 2006.

In 2006, the US government summarized the consequences of Venezuelan oil production being affected.

> A model developed for the Department of Energy estimates that a 6-month disruption of crude oil with a temporary loss of up to 2.2 million barrels per day—about the size of the loss during the Venezuelan strike—would, all else remaining equal, result in a significant increase in crude oil prices and lead to a reduction of up to $23 billion in U.S. gross domestic product. A Venezuelan oil embargo against the United States would increase consumer prices for petroleum products in the short-term because U.S. oil refiners would experience higher costs getting replacement supplies. A shutdown of Venezuela's wholly-owned U.S. refineries would increase petroleum product prices until closed refineries were reopened or new sources were brought on line. These disruptions would also seriously hurt the heavily oil-dependent Venezuelan economy.
>
> U.S. government programs and activities to ensure a reliable supply of oil from Venezuela have been discontinued, but the U.S. government has options to mitigate short-term oil disruptions. For example, activities under a U.S.–Venezuela oil technology and information exchange agreement were stopped in 2003, in part, as a result of diplomatic decisions. In recent years, U.S. oil companies have not sought assistance from the U.S. government with issues in Venezuela because the companies do not believe that federal agency intervention would be helpful at this time. To mitigate short-term oil supply disruptions, the U.S. government could attempt to get oil-producing nations to increase their production to the extent possible, or could release oil from the U.S. Strategic Petroleum Reserve. While these options can mitigate short-term oil supply disruptions, long-term reductions in Venezuela's oil production and exports are a concern for U.S. energy security, especially in light of current tight supply and demand conditions in the world oil market. If Venezuela fails to maintain or expand its current level of production, the world oil market may become even tighter than it is now, putting further pressure on both the level and volatility of energy prices.

What's more, those effects would be a lot more profound today once you consider that America's economy is set for a very slow recovery after averting a full-fledged depression, and that oil markets are increasingly getting tighter.

Negroponte on that scenario:

> I doubt you that that will happen. You would have to cut trade. And we're going to do that? To cut imports from Venezuela and jeopardize investments of large American corporations?

Indeed, few expected the United States to pursue more sanctions, lest while confronting Iran. Pumphrey, the former DOE official and current head analyst of the CSIS, was also skeptical days ahead of the decision.

> It would be a high bar. But if they had conclusive proof that there was support for terrorism, they would impose sanctions. We've imposed sanctions to important oil producers before. It drives the oil industry crazy, but we would move toward that. But the bar would be high to show its threat to us.
>
> Initial sanctions would be financial flows to the government. You don't have to ban oil imports. You can take financial sanctions before trade. It doesn't lead necessarily to that. As if you get into financial flows you interrupt PDVSA ability to process money.

It's significant that the United States didn't target PDVSA or Chávez's over terrorism concerns. It did it over a legal shipment of fuel additives. But even if it's just playing to a domestic audience, sanctions hurt Venezuela's oil plans and empower Chávez.

Latin American governments, including Chávez, are wondering whether the Obama Administration is building a case against Venezuela. Brazil is alarmed. In September 2011, the Treasury Department imposed sanctions on four Venezuelans officials for "providing arms and security to the FARC," one major general, one congressmen of Chávez's party, the alternate president of the Latin American Parliament, and an intelligence officer. Their assets were frozen and Americans are prohibited from doing business with them.[11] As already analyzed, two more Venezuelan citizens were also sanctioned for supporting Hezbollah, along with a Venezuelan bank subsidiary

[11]www.treasury.gov/press-center/press-releases/Pages/tg1295.aspx

of an Iranian bank.[12] They join a list of four high-ranking Venezuelan military officials already sanctioned in 2008 for ties with the FARC.

Chávez response was swift: "I dare President Obama to present the proof of this infamy . . . that are part of the aggressions toward Venezuela, of the Yankee empire's intention of including Venezuela one day in the list of failed states or in the list of countries that support terrorist. As a terrorist country," he said to Venezuelan media.

In fact, Chávez named one of those targeted by the Treasury Department, General Henry Rangel, as defense minister in January 2012, not only defying Washington, but also giving power to the most anti-American segment of the Venezuelan Armed Forces. Days later, a White House spokesman said the "concerns about Rangel Silva are well known and of long standing. In 2008, the Department of Treasury designated Rangel Silva under the Kingpin Act for materially assisting the narcotics trafficking activities of the FARC."

The Obama administration in January 2012 moved from economic sanctions to more headline grabbing diplomacy. Shortly before Iran's Ahmadinejad started a four-day tour of Latin America in January 2012, the State Department expelled Venezuela's consul in Miami. The diplomat was caught on tape in 2008 discussing with Mexican computer hackers how to infiltrate sensitive US websites, including those of the Pentagon, White House, and nuclear plants, in what would appear to have been an Iranian plot that also involved Cuba.

No charges were filed. Chávez brushed off the conspiracy theory, and US security was never threatened. It also does seem highly unlikely that any of the three countries allegedly involved have any cyber-warfare capability to threaten American interests with the help of Mexican hackers.

"This decision makes it even more apparent that Washington is unwilling or unable to implement policies that will constructively influence Latin America," said an analysis published by the Washington-based Council on Hemispheric Affairs, a think tank, about the consul's expulsion. The spat is useful though as "votes could be picked up from rightist exile groups (in the US) coming from Cuba, Nicaragua, and Venezuela among other disaffected conservatives, who hold passionate thoughts on their personal political agendas."

---

[12]http://foreignaffairs.house.gov/112/szu062411.pdf

Whatever the case, Obama is hard pressed to come up with new policies toward the region to adapt to new circumstances. The status quo continues to backfire and America's interests are further at stake, especially in regard to energy. Both the United States and Venezuela will hold presidential election in the last quarter of 2012. For political reasons, both Obama and Chávez will be likely tempted in this last year ahead of their reelection campaign to seek political gains with voters, at the expense of a more rational foreign policy. Confront Chávez and Castro, interests groups will say in the United States. The evil empire is out to get us, Chávez will reply. And the relations of both are likely to suffer.

And in the meantime, Latin American policy remains in disarray. Valenzuela, the top diplomat for Latin America, left his post in August 2011 to return to academia without leaving much of a mark, other than the usual troubled relations with Argentina and the Chavista block. Obama decided to avoid a battle in Congress and nominated little known Deputy Assistant Secretary for Hemispheric Affairs Roberta Jacobson to fill the job. Of the Clinton idealist school, she has served in midlevel posts and as deputy chief of the embassy in Uruguay, and like her predecessors she has echoed more of the rehashed status quo of containing Chávez, immigration, the drug war, and monitoring democratic rule in the continent.

The FTA standoff

Former President Alvaro Uribe bet a big part of his eight-year legacy and Colombia's future on a steadfast partnership with the United States. His plan was to expand on a longstanding diplomatic and military alliance by deepening economic ties through an FTA. Colombia, flanked by Venezuela and Ecuador, was ready to sacrifice economic relations with its neighbors if it came to it, but the FTA would make up for any diplomatic or economic losses. Colombia would be the success story of the US-backed economic model and neoconservative ideology, a vehicle to balance the regional shift toward the left and a counterweight to Chávez's Bolivarianism.

In 2009, Uribe agreed to give US forces limited access to seven military bases, including one near Venezuela's border, as well as basing a small number of troops in Colombia. The US Air Force, in a draft budget proposal presented to Congress, said that investment earmarked as part of the increased presence in Colombia would offer a "unique opportunity for full spectrum operations throughout South America" in a series of scenarios, including threats from "narcotic funded terrorist insurgencies and anti-US governments," a reference that although edited out in the final proposal was interpreted by most South American countries, including powerhouse Brazil, as a potential regional military threat.[13] For many, including Brazil, the agreement with Colombia gave a blanket permission to eavesdrop on all countries, but Chávez described it as a smoking gun indicating that the United States intended to use Colombia to remove him from power, the ultimate leader of the anti-US movement. Colombian courts eventually found the deal unconstitutional.

Colombia delivered what Bush and Obama asked for, but got little more than weapons in return, and to that effect, a very small percentage of its defense expenditure. Uribe pleaded for a last minute Obama concession to support the FTA. But Obama held out, a domestic political calculation that would prove disastrous for America's interests, not for economic or trade considerations, but strategically. It was in effect Colombia's last straw.

The question is not whether Colombia needed to improve its record on human and labor rights and prosecute those responsible for gross violations over the past few decades, including high-ranking military and former and current government officials, legislators, and judges.

But is it in America's interest to be the moral overlord that imposes that? And does it even accomplish anything? Colombia indeed improved its human rights record and is now more aggressively prosecuting corrupt officials and human rights violations, but not because Democrats in the United States demanded it. Colombia is transitioning, not unlike other countries in the region, and its decision to tackle its human rights record is domestically driven, in the same way that it has committed to the most ambitious poverty-cutting goal. To put it simply, Colombia needed to reclaim the military initiative and human rights issues were ignored, but Washington's decision to immerse itself in the process as a supervisor was counterproductive to American interests and to Colombian efforts to pacify the country and prosecute those responsible.

Democrat idealists miscalculated Colombia's political and economic evolution. Obama's policies arguably made it harder for Colombia to strengthen its economy, a prerequisite to fight drug and guerrilla armies while strengthening its institutional stability. At best, they interfered in

---

[13]Department of the Air Force, Military Construction Program. Fiscal Year 2010. Budget Estimates. Justification Data Submitted to Congress, May 2009.

Colombia's struggle to extricate itself from 50 years of continuous war and institutional weakness.

On the economic front, the FTA with the United States will only come into force sometime in 2012, but in the meantime Colombia implemented other FTAs with Brazil, Argentina, Uruguay, Paraguay, and Canada. An FTA signed with the European Union will also likely come into effect in 2012, and talks have begun with South Korea and Japan. That means American exports to Colombia, worth $12 billion in 2010, will be increasingly less competitive until it implements its own agreement. In 2011, bilateral trade was more than $37 billion.

A Santos doctrine?

Uribe couldn't defend politically unbalanced relations with the United States. But with little else to lose and looking forward to a third reelection campaign, he stuck to unwavering support for Washington's policies. Colombia was diplomatically isolated in Latin America, compounding its economic woes and undermining its antiterrorist war against the FARC.

With just a few days left in power, Uribe pretty much rigged a diplomatic time bomb with Venezuela and the rest of Latin America. He showed alleged proof of FARC presence in Venezuela, including photos and specific coordinates of five guerrilla camps a couple of dozen kilometers from the border. Obama demanded investigations and Venezuela replied that the proof was fabricated. In another symbolic move, Uribe—representing himself, not Colombia—denounced Chávez's support for guerrillas in the International Criminal Court and in the Inter-American Court of Human Rights.

At that point, in August of 2010, Juan Manuel Santos took office. At stake was an embargo imposed by Chávez on Colombian imports in 2009 when relations with Colombia were at historic lows. Bilateral trade of nearly $7 billion in 2008 (or about 20 percent of the country's total) dropped to $4 billion in 2009 and to $1.5 billion in 2010. Trade was reactivated only after Santos had made peace and Venezuela agreed to pay back more than $1 billion in pending payments claimed by Colombia.

A second motivation was the war on terrorism. Colombia has been in a state of war for nearly half a century, fighting several insurgencies, paramilitary armies, drug mafias, and the combinations thereof. The biggest is the FARC, with 8,000 combatants spread across Colombia. It's a well-armed, well-trained army financed through the drug trade. By the end of the twentieth century, it had pushed Colombian security forces and the state out of large swaths of Colombia, at one point nearly encircling the capital Bogota. Lawlessness in the border regions of all neighbors was allowing the FARC vital breathing room.

Santos quickly moved to completely overhaul the country's foreign policy. It was urgent. He had inherited what amounted to a trade embargo with Venezuela, saber rattling in Ecuador, few friends in the rest of Latin America, and only conditional support from the United States. There was talk of a regional arms race and warmongering. Colombian exporters were losing billions every year and the FARC were exploiting bilateral frictions to their advantage.

Santos's turnaround was remarkable. Only three days after taking office, he met Chávez in Colombia and reestablished relations. The Venezuelan leader agreed to deny Colombian insurgents use of his territory and to seek mechanisms to reactivate commerce. Colombia agreed not to interfere in Venezuela affairs. Both leaders also agreed to jointly develop infrastructure projects of mutual benefit, which for Venezuela ultimately means pipelines for its oil to reach the Pacific Ocean.

Santos took the next step. Before his first month in office was over, he quietly decided not to seek ratification of the military deal with the United States that he had drafted as defense minister and that Uribe had signed. The country's highest court days before had denied the government authority to implement the agreement because it hadn't been approved by Congress. Santos could have easily pushed it through with the overwhelming majority his supporters hold, but he didn't. And he can resurrect it when he chooses, thus giving him a bargaining chip. Santos's decision was interpreted as a carrot to reestablish relations.

Santos visited Chávez in Caracas in November 2010, where more agreements were signed, including cooperating in the wars against drugs and terrorism. Few gave credibility to warming relations, much less to Chávez's willingness to take on the FARC. Chávez didn't deny Colombian guerrillas were using Venezuelan territory, but he rejected government complicity. The coordinates Uribe delivered with exact locations of camps turned out empty, which doesn't mean those camps didn't exist in the past. The deal with Chávez was that Colombia would hand over information to capture guerrilla members and Venezuela would act upon it. Colombia had nothing to lose.

By the end of November 2010, Colombia also reestablished relations with Ecuador, a Bolivarian country, with the mediation of Chávez. By then Santos had visited Brazil and Argentina, mending tense ties with the two. In less than four months, Colombia had turned back almost a decade of diplomatic setbacks with the rest of Latin America.

It was time for Chávez to reciprocate. Venezuela started paying back its debt to Colombian exporters of around $1.4 billion, and took concrete steps to reactivate bilateral trade more vital to Colombia than to Venezuela. The task was difficult because Venezuela had decided to opt out of an FTA with other Andean nations. There was room for significant improvements, but it would take months, if not years, for Colombian exports to once again reach almost $7 billion without a legal framework in place. Both countries continue to search for alternatives. Meanwhile both Santos and Chávez often boasted of their warm ties, baffling observers. During one interview with CNN in February 2011, Santos said "the mood in Latin America changed 180 degrees" thanks to reestablished relations between Colombia, Venezuela and Ecuador.

In late March, when the strategic shift was well in motion and critical voices surfaced in the United States and Colombia, Santos made an unapologetic defense on his relations with Chávez before the Colombian-American Chamber of Commerce.

> We have good relations with our neighbors, not because I've become Chavista. Nor do I think Chávez has become Santista. It's because it's mutually beneficial, because we share a 2,200 kilometer border, and if we respect our differences and collaborate in our common interests, everyone will win. What's happened with Venezuela . . . has been a net gain for Colombia in everything. We are well underway and we also want to remain close to the US, which has been our strategic ally. There are things that are not at all exclusive. People believe, for example, that our new relations with Venezuela and Ecuador will put Colombia at odds with the US. It's exactly the opposite. Colombia can be a lot more constructive in relation with the region if Colombia has good relations with the region because we are strategic partners.[14]

A week later Santos would meet Obama in the White House, and two days later Chávez in Colombia, but not before he faced the biggest test to his

---

[14]http://wsp.presidencia.gov.co/Prensa/2011/Marzo/Paginas/20110330_07.aspx

foreign policy turnaround. Walid Makled, a Venezuelan of Syrian descent and labeled by US authorities as one of the world's top drug kingpins, had been arrested in Colombia in August 2010. In several media interviews in jail, Makled said he could document payoffs to dozens of Venezuelan military officers, officials, and Chavista politicians; proof, he said, of the government's direct involvement in the drug trade. He offered to cooperate with American authorities. But after intense diplomatic pressure from both Washington and Caracas, Santos decided to hand Makled over to Chávez, a decision heavily criticized by idealists both in the Democratic and Republican parties, and by Uribe in Colombia. Santos explained that under Colombian law Venezuela had priority because it was the first to request the extradition.

Uribe's former Vice President Francisco Santos in an interview questioned the decision.

> It's a mistake to underestimate the importance of the United States. That said, it's also a mistake to overestimate it. The US is our most important strategic partner. We share fundamental values, but Colombia has been mistreated by some sectors in the US.
>
> [Juan Manuel] Santos is seeking equilibrium, but that's a dangerous game.

Indeed, it was an uncomfortable decision for President Santos. The confirmation that Makled would be extradited to Venezuela came a day ahead of his first official visit to the White House, when Obama finally confirmed his support for the FTA, and in the same week that Santos further strengthened his new ties to Chávez with the signing of 16 bilateral agreements. Colombia got an extension of trade mechanisms favoring its exports, while some kind of new bilateral trade agreement was drafted. And Chávez resurrected his long-held dream of building an oil pipeline from Venezuela, through Colombia, to the Pacific Ocean.

American idealists were reactionary, claiming Obama had lost Colombia to Chavismo. But nothing was further from the truth. Indeed, the pragmatic Santos Doctrine was showing results. In one same week, Colombia secured Obama's support for the FTA, trade benefits for Colombian exports to Venezuela, and more importantly, it turned out, Chávez's first real efforts to deny Colombian insurgents safe haven.

Two weeks after Colombia agreed to extradite Makled, at the end of April 2011, Venezuelan police arrested Joaquin Pérez, the FARC's representative in Europe, as soon as he arrived in Caracas. Santos personally called Chávez to request his arrest. He was deported two days later to Colombia. Chávez came under intense pressure from left-wing political groupings inside and outside Venezuela, which he brushed off. He had already handed over ten other people requested by Colombia, but Pérez was the highest ranking FARC militant until then. One month later, late in May, Chávez outdid himself. Guillermo Torres, a high-ranking guerrilla commander, was captured on a Colombian tip. Chávez personally promised to extradite him, again defying vocal criticism from his own coalition, especially the Venezuelan Communist Party. Torres's case remains stuck in courts.

Santos has defended Chávez and repeatedly said that the Venezuelan leader is upholding his end of the deal and cooperating in the war against drug and illegal armed groups. Santos has said relations with Venezuela are of "great strategic importance." Santos often talks to Chávez over the phone and both say efforts to wrench their rapprochement will not work. Relations with Ecuador have also improved remarkably. In September, the government of Correa used its executive power to block attempts to issue an international arrest requested by an Ecuadorian court against five of Colombia's former and current top ranking military and police officers for their alleged role in the cross-border raid in 2008.

Former Vice President Francisco Santos is wary of the diplomatic U-turn his cousin is leading, especially when it comes to trusting Chávez:

> It's the right strategy in the sense that it shows that Colombia is doing all it can to improve relations. It comes at a high cost though in the sense that an authoritarian leader in a democratic region doesn't get punished. In doing so, Colombia joins the regional vogue of gauging [national] interest a different way, disregarding democratic institutions. But that's international relations for you. You don't have friends, but interests.
>
> The heart of the problem is that [Venezuela] has a pending issue with us, which are the FARC and ELN. They are comfortable over there. It's their rearguard. It's a problem that President Santos has not addressed, which he hasn't insisted on, but it's a real and ongoing problem.
>
> Chávez also thinks the FARC are his rearguard for a perceived US invasion. They are a resource that he uses against Colombia, like Iran with Hezbollah, like a pawn. It's an ideologically fluid relation. The

FARC is allowed to rearm, to heal, and the collusion is articulated in the drug business. There is real support. I don't have the slightest doubt. I think the US will eventually include it in a list of state sponsors of terrorism. The evidence supports the case. The only reason they haven't already is oil.

But President Santos is steadfast in his diplomatic shift, despite intense political pressure from Uribe, his predecessor and ally, and others who criticize his approach. High level bilateral talks continue, even at the military level. In an interview with the Washington Post's senior associate editor Lally Weymouth in late August 2011, he reviewed his strategic shift in foreign relations and their impact. Here are a few excerpts that explain the Santos Doctrine:

> I personally have always been a good friend of the U.S. I am and will continue to be. I was educated there and for me, having strategic good relations with the U.S. is of the utmost importance. But I think they are not mutually exclusive, to have good relations with your neighbors and with the U.S.
>
> Yes, [relation with Venezuela] were not almost at an end; they were at an end. We had no diplomatic relations, no trade. The only discussions we had were through the media and we were talking about war, which is inconceivable. So I decided to have cordial and cooperative relations not just with Venezuela, but also with Ecuador, a country with which we also had no relations. What we decided is that it is in our interest to have a region that is not on the verge of war.
>
> [Trade relation] will never go back with the regime that we have in Venezuela. They don't believe in free trade; it's as simple as that. I negotiated the free trade agreement 20 years ago with Venezuela. The trade went from $300 million to $7 billion and then it went back to zero because we had no trade. We are restoring it but on a different basis because they don't trade freely.
>
> Fortunately, the financial crisis of 2008 did not hit us as hard as other countries. Since 10 years ago—when we had our big crisis—we started to develop ways to protect ourselves from international financial crises. Our financial system was very strong in 2008 and today is even stronger. We are protecting ourselves with monetary measures and by strengthening our financial system and by starting to speak to the region to see how we can confront this crisis multilaterally.

The ministers of finance of the region got together and decided to strengthen the regional financial institutions, like the Corporación Andina de Fomento [Andean Development Corporation]. They also decided to strengthen the regional reserve funds, and to start talking to each other and the central banks to see what type of additional measures could be taken to protect the region, which is growing—in contrast to the U.S. and Europe. We are exploring if we can increase intraregional trade, because we are afraid that demand from the traditional consuming countries like Europe and the U.S. might come down.

I hope we don't have a double dip.

[The US is Colombia's top trading partner] and I hope you will continue to be. That's why I am so interested in the approval of the free trade agreement.

I have my fingers crossed. The U.S. has been our most important trade partner and our most important investor. It is not only in the interest of Colombia—it is in the interest of the U.S. that Colombia has free trade. We are a source of growth for the U.S. We are 46 million Colombians. We are growing at a relatively high rate; we are taking millions of Colombians out of poverty. We want to lower the poverty rate by at least 7 or 8 percent.

The security situation is under control. We still have security problems because we still have an interior conflict. But the FARC is weakened; they are politically defeated, but they still have enough muscle to commit acts of terrorism and that's what they have been doing.

[Chávez] has made a commitment to me that he will not allow the FARC to use his territory as a sanctuary.

So far he has given me no grounds to say he is not complying. There probably are guerillas in Venezuela, but he says it is without his say-so and that if I pinpoint where they are, he will go and get them. On two different occasions, I have pinpointed and he has delivered.[15]

By the end of November 2011, Venezuela had signed a bilateral trade deal eliminating duties on more than 90 percent of Colombia's exports. A

---

[15]www.washingtonpost.com/opinions/an-interview-with-colombian-president-juan-manuel-santos/2011/08/24/gIQADrqFgJ_story.html

month later, just before the year's end, Colombia extended a natural gas export deal vital to Venezuela. Chávez and Santos also reaffirmed their commitment to building the oil pipeline. Chávez in fact said in February 2012 that studies were under way for the "three-way" or multilateral project, referring to Chinese investment, and possibly from other Asian countries like Japan. Even if the construction is ultimately uncertain, like so many of Chávez's projects, it makes sense Colombia would support the pipeline to gain more leverage over Chávez, and more bilateral stability vital for Colombia.

Santos's doctrine almost takes a literal interpretation of Sun-Tzu's Art of War advice: "Keep your friends close, and your enemies closer." Both presidents are on opposite sides of the ideological spectrum, but Santos—a realist—has in effect co-opted Chávez to the benefit of Colombia, at least temporarily. Naturally, idealists will say, the Venezuelan *caudillo* is volatile and prone to sudden changes and correction. Chávez could be using the window of good relations to clean up his image ahead of elections, and in the future revert to more hostile policies.

A realist assessment though suggests otherwise. Colombia's former intelligence chief and former BP advisor Peñate analyzed bilateral relation:

> Colombia and Venezuela don't have brotherly ties. They are more like Siamese. You can fight with a brother, but you can't separate from a Siamese. We are integrated. More than 2 million Colombians live over there and our border region is active. Our cultural ties run deep.
>
> As much as Colombia depends on Venezuela, Venezuela depends on Colombia, especially for basic food staples. It's hard for Chávez to keep hostile relations with Colombia. Fighting with Colombia is not popular in Venezuela. And Chávez is not Gaddafi. He cares about public opinion. His popularity is the check and balance to his power and relations concern Venezuelans.

To be sure, regardless of what Chávez does in the future, Colombia has already won. Bilateral ties historically have been mostly one-sided in favor of Venezuela. But Venezuela's economic prosperity, as conceived by Chávez, transits through Colombia. Chávez needs to increase oil production and to send it to Asia, and Colombia is critical to both goals. Venezuela needs natural gas from Colombia and the pipeline. By intertwining the destinies of Colombia and Venezuela through bilateral commerce, gas exports, or pipeline interconnections, Santos seeks to balance relations, to make it more costly for Chávez to threaten Colombia in the future.

On the terrorism front, even if Chávez wanted to revert to helping the FARC, it's very unlikely the guerrilla group would ever trust his intentions again, as its role as pawn in relations with Colombia's government has been exposed.

Santos already won the moral upper hand too. If Chávez decides to break the deal after the elections or further in the future, his standing in the rest of Latin America would suffer. Santos has unquestionably delivered on concrete measures to improve relations, including Chávez's top concern involving the military agreement with the United States. Latin America, starting with Brazil, would isolate Chávez if he unilaterally broke with Santos.

Furthermore, Chávez at this juncture probably has more at stake than in the past. Domestically, his popularity would suffer. A majority of Venezuelans have embraced new relations, especially those in the center that will increasingly determine electoral outcomes in the country. Economically, Venezuela needs Colombian imports simply because replacing them has been painfully expensive. And militarily, Colombia can always quickly pass the military agreement with the United States.

That is precisely Santos's Doctrine. By confronting Chávez, Colombia and the United States lost almost all leverage over him. Colombia's security was seriously threatened more by the break in relations under Uribe than by Chávez's actions. At the height of the bilateral impasse, Chávez had little at stake and Colombia stood to lose all security gains as long as Venezuela ignored the FARC presence in its territory. Santos correctly concluded that Colombia's best interests would be served by balancing relations and mutual interests with Venezuela, while strengthening ties to the rest of Latin America and the United States.

Former US intelligence chief Negroponte offers an American perspective:

> Santos is being more correct in relations. He's trying to manage potential confrontation. But I don't think he's altered his view that these are two quite different political systems.

Sometime you have to make a pact with the devil to achieve maximum security. These are sometimes difficult trade-offs, but understandable if you can accomplish something to strengthen the security of Colombia in the future and to bring the narcotic problem under control.

Roosevelt dealt with Stalin to deal with Hitler. Trust but verify, Reagan used to say. You have to be careful how you deal with people like Chávez. You have to make sure you make sound arrangements and on a step by step basis.

But if you ask me about strategic trust, I wouldn't place strategic trust on Hugo Chávez.

9 THE WAY FORWARD

In the December 2011 interview with Venezuelan media, Obama reviewed relations with Latin America, a disappointing rehash of his good intentions. And in criticizing Venezuela's government, he insisted on prioritizing security, not the economic component of relations with the region. His comments increased diplomatic tensions with Latin America ahead of the Summit of the Americas in April 2012 in Colombia.

Obama said:

> We're bound by shared values, a common heritage and common interests. With no other region of the world does the United States have so many connections. That includes the tens of millions of Hispanic Americans across the United States.
>
> As President, I've committed the United States to a new era of partnership with the region based on equality, shared responsibility, mutual interests and mutual respect.
>
> Some in the region—including the Venezuelan government—have demonstrated anti-American tendencies. But to be blunt, I don't think the people of the Americas want to live in the past because they're interested in the future.
>
> My approach to Venezuela is guided by my policy toward the Americas as a whole. The United States seeks to engage on issues of common interest with governments that are interested in working constructively with us. For example, the American people—and I believe the people of Venezuela—have an expectation that their governments can and should cooperate on matters of mutual concern, including what should be common struggles to prevent terrorism and drug trafficking.

> For their own reasons, Venezuelan authorities to date have shown little interest in that kind of cooperation. And, as I said, we're concerned about the government's actions which have restricted the universal rights of the Venezuelan people, threatened basic democratic values, and failed to contribute to the security in the region.

Obama's reading of Latin America reality is off. He's right to say that most people in the region, policymakers and citizens alike, are tired of outdated ideological standoffs, and that includes listening to Chávez rant about imperialism or socialism. But for exactly the same reason, they are also tired of the United States preaching about democracy and Cuba. Put simply, most in Latin America wants Washington to drop its moral, idealist policies and to join the Brazil-led consensus.

The US isolation in the region was exposed during the Cartagena summit. Obama found few friends, while South American countries, led by Colombia and Brazil, were unable to convince Washington of the need to revise its outdated policies on Cuba, on the drug war, on immigration, and on free trade. There was no final declaration, an outcome that highlights Latin American assertiveness and US continental isolation. Chavista countries kept mostly quiet and savored the diplomatic loss of Obama.

South America's rising assertiveness was further illustrated with the Argentinean government expropriation in April 2012 of a 51 percent stake in the country's biggest oil company Yacimientos Petrolíferos Fiscales (YPF). The YPF stake was owned by Spain's Repsol and was worth as much as $10.5 billion, although tribunals will set the ticket price. Bolivia in May 2012 also expropiated the country's power distribution company owned by a Spanish firm.

The decisions initially raised alarms that another wave of Chávez-like statist expropiations was taking hold in the region. But it subsided within weeks as other Latin American countries distanced themselves, and as Western companies, the IMF, and other global institutions with vested interests in the region concluded that the expropriations were an exception, not a trend.

The expropriations are not a symptom of Latin America's instability, but of just the opposite. Argentina feels it is powerful enough to challenge Spain on YPF, and the United Kingdom on the Falkland Islands. European countries are in decline, while countries like Argentina are on the rise and demanding their place in world affairs, like Brazil and Venezuela have already, and Colombia is likely to do in coming years.

Indeed, the United States challenged Spain in 1898 to debunk Spain as a superpower. The expropriations are the modern equivalent that speaks to Spanish decline and the assent of Latin American countries.

But how should the US deal with Hugo Chávez? What should relations with Latin America be like? The question has been baffling policymakers for over a decade. In his first term, Bush "the second" tried direct confrontation, including openly and actively propping up the opposition fighting to depose him. While no overt involvement or prior knowledge of the 2002 military coup has been proven, Bush administration officials at the very least were ambiguous in their defense of a democratically elected government.

Relations with Venezuela and the region have arguably worsened under the Obama administration. Indeed, Republicans and Democrats have failed to successfully address the Chávez's conundrum for the simple reason that they have all positioned themselves as ideological antagonists. This benefits Chávez, who put anti-Americanism at the heart of his regional policy. Venezuela intentionally has conditioned relations to be perpetually at odds with the United States.

Conventional wisdom is that there is little room for a policy change toward Chávez.

Former President Uribe:

> When countries sacrifice their democratic principles for economic benefits they end up without economic benefits and without democratic principles. For me, democratic principles are above any economic interest.
>
> Stalin used to say capitalism can be destroyed easily because capitalism thinks only about capital benefits, that it's willing to bend its values. That can't be done. The truth is I have a very clear point of view on this, which is why I don't like these questions.

Former US intelligence chief Negroponte:

> We shouldn't embrace Chávez. But we should maintain a correct relationship with the state of Venezuela because we have a variety of shared interests. We have commercial and economic ties. We just don't have a warm and fussy relation with Mr. Chávez himself.
>
> I don't think we want to make a gratuitous enemy out of Venezuela. That would be a mistake. We need to try to manage that relationship in as low key and constructive way as possible.

The problem is that the status quo is hurting the United States more than Chávez, plain and simple. The United States will not benefit from escalating the standoff unless it has very strong reason to do so simply because it doesn't want to deal with replacing 10 percent of its oil and products import demand, certainly not until the Iranian standoff is resolved. Obama could do it, but the oil market is tight as it is to add a new element of price volatility.

Both Venezuela and the United States are moving to decrease their mutual dependence anyway, in the US case to gain more leverage and in Chávez's case to decrease his exposure. The United States is very likely to increase imports from Canada into the future, while Venezuela is maneuvering to diversify export markets toward China and Asia.

The realist conclusion is that the United States currently has no leverage on Venezuela. And this is unlikely to change any time soon. That means that confronting Chávez directly—especially on ideological grounds—will inevitably fail.

Weisbrot, of the CEPR, defends direct talks instead:

> The policy of going after Chávez has failed. The Bush administration wanted to isolate Chávez. What did they do? They isolated themselves. It's the worst relations with Latin America that they ever had. In contrast, Chávez's relations with Brazil and Argentina are very close.
>
> Chávez doesn't have to answer to others on foreign policy. It's elections that he has to win. And like in the US, foreign policy is not what voters care about, but domestic policy.
>
> Obama knows what to do. He's not doing it because he's not paying attention. The White House is not involved.
>
> Obama should normalize relations. He could do it even easier than President Santos. And that would be the end of the problem.

Playing nice will not work because Chávez by definition will always be at odds with the United States. And in the idealist battlefield, the Venezuelan leader has the upper hand simply because he is able to project more regional power than the United States is prepared to. That is why he appears to be winning.

Pumphrey:

> I don't know of any other tactic [to deal with Chávez] other than the one we have going forward. Try not to do things that legitimize him by confronting him. It's a frustrating process. He's very clever. I witnessed in the Bush administration how they tried to counter each blow, and that just ended up strengthening him. We need to do what we can to work with him without sacrificing our values. It's a mushy answer but it's a mushy situation.

Former Colombian Vice President Santos concurs:

> I don't think you need a policy to neutralize Chávez. We did that by reestablishing relations. The rest is developing your interests, your markets, as part of a strategy that goes beyond Chávez. Our interest is not to weaken or strengthen him. I do think it's in our interest to have a change over there, to recover democratic institutions and the economic situation which is unraveling. Besides, Chavismo expansionism climaxed. It has been a disaster and people are running away from it. Wealth destruction in Venezuela is huge.

The only way to deal with Chávez is to hit him where it hurts, without hurting American interests in the process. America does not need another Iran just across the Caribbean. Instead, the United States must target Chávez's oil immunity to its advantage, and that means implementing policies to increase production, not decrease it through sanctions and diplomatic pressure.

His power domestically and abroad is based on oil. He is more dependent on oil to perpetuate his power than the Iranian or Saudi regimes, for example, which have other authoritarian resources at their disposal. Chávez doesn't have an army able to suppress its own people, nor does he enjoy enough popular support or internal unity to get away with it. Perhaps the closest comparison is Russia, which also has a fractured political system and depends on oil to bankroll political and economic stability.

On the surface, it might appear that Chávez's regime is more stable than its predecessors, but it's not. Chávez is more exposed than any previous government or regime in Venezuela to low oil prices. No system of government can survive in Venezuela without enough oil revenue. That is how he came to power in 1999, after almost a decade of low oil prices that battered the economy and democratic institutions. Chávez has only made his regime even more dependent than previous governments by significantly increasing public spending.

During his 12 years in power, he has defended high oil prices, and indeed benefits from the tight oil market and Middle Eastern instability. But Chávez is in much worse shape than Iran when it comes to the oil industry. He has no room to manipulate oil production. Venezuela has a lot of potential, but not enough capital or know-how. That means that in order to pay for his Revolution, Chávez has no choice but to pump at maximum capacity for years to come. And that is precisely what America wants: as much Venezuelan oil as can be possibly placed on the market.

Roger Tissot is an independent energy consultant based in Canada specializing in Latin America's energy with over 15 years of experience, which includes directing Latin American analysis at PFC Energy, a global energy consultancy. "If re-engaging with foreign oil companies is necessary to his political survival, then Chávez will do it," the *New York Times* quoted Tissot as saying. "He is a military man who understands losing a battle to win the war."

Chávez has already cornered himself. To survive internationally and project abroad he needs more oil because he needs more cash. The more oil he brings online, the more bearish pressure he puts on oil prices, cyclically increasing the need to boost production even further, and so on. This is not true of every country, of course, not just because most have fully developed their industry, but because their economies are in better shape to survive low oil prices.

Andrés Peñate, the former energy strategist and head of Colombian intelligence services, explains Chávez's Achilles' Heel:

> Chávez will give up power only through what brought him there in the first place: the barrel of oil. Venezuela is an oil company with armed forces, an anthem, and coat of arms. PDVSA and Venezuela need an oil price threshold to survive. And all Chávez has accomplished is to increase that threshold. He adds programs that don't add revenue, and expands them to Latin America and even the US.
>
> The only thing that can remove him from power is if oil prices fall to such a level that Venezuelans second doubt the government's ability to deliver on its commitments, including the cheapest gasoline in the world, free electricity, free healthcare, free tolls, etc.

How low do prices have to go and for how long before Chávez feels the pressure? Estimates vary with time, not only because Chávez spends more, but because he increases his debt, while limiting Venezuela's ability to capture high oil prices by subsidizing it for others. Estimates of Venezuela's breakeven oil price for 2011 vary and mostly fall within a range that starts at $80 a barrel and goes as high as $110 a barrel. Figure 9.1 shows the breakeven costs of several major oil producers. Notice Venezuela's are the highest. But if prices were to stay at around $80 for an extended period, Chávez's hold on power would be seriously threatened.

US policies geared to increase production in Venezuela could be a modern and much more discreet application of the strategy Reagan used on the Soviets to win the Cold War, ushering an era of low oil prices and US supremacy. Major oil producers can only really be threatened when oil revenues are targeted, as the case of Iraq under Saddam Hussein illustrates. In Iran's case, the United States resorted to directly targeting the country's ability to export oil because it's an urgent matter and because it's a last resort, as Obama described it himself. But targeting Venezuela's oil exports would be suicidal, on top of disproportional, meaning that the best option for the United States is to seek a long-term strategy to lower international oil prices, much like Carter and Reagan did to defeat the Soviet Union.

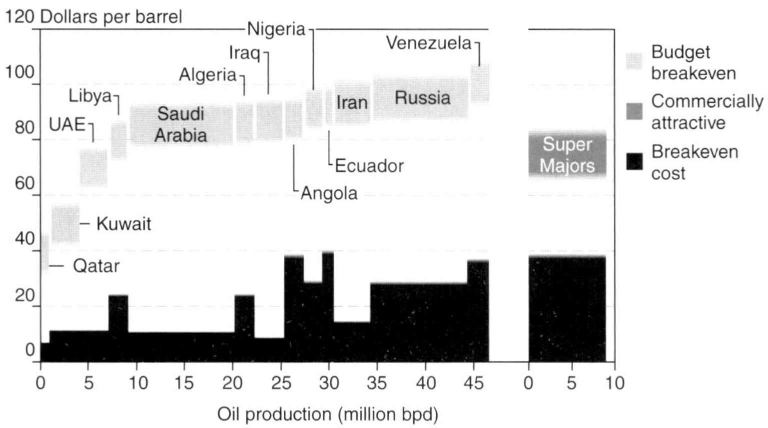

FIGURE 9.1 Oil price requirement for selected producers
Source: IEA WEO 2011

Here is Negroponte's analysis:

> When I was director of national intelligence, I used to say that the higher the price of oil the longer a person like Chávez can remain in power. There's a point where lower oil prices would really cramp his style, like subsidizing Cuba.

A new approach

Geopolitically speaking, the United States beyond a doubt has more important issues to deal with than Latin America. That includes Chávez's Bolivarianism. But the status quo of handling regional policy as a second-tier priority at best, to be mismanaged by bipartisan idealists who exploit nonvital issues for petty domestic gains, is undermining America's economy indirectly by putting bullish pressure on oil prices.

America's blind spot can be addressed by implementing a realist policy south of the border, which will improve the country's standing in the region while simultaneously undermining anti-American forces. More importantly, a new approach would likely foster more oil production from the region, which at this point in time is critical to shield the United States and its OECD allies from another recession that could seriously threaten security.

Ultimately, the only way to boost oil output out of Latin America is through more private investment. The US government can't do much because it will only meddle into corporate decisions as a last resort, unlike European and Asian national champions with majority government shares that can be more easily influenced. Pumphrey on America's options:

> I don't know what it would do. I spent eight years during the 1990s and early last decade trying to foster energy policy toward Latin America. There's been a lot over time of telling Latin American countries to produce more. But what do we bring to the table as incentives? We don't have financial resources. The concept is out there, but how to make it operational is difficult. We tried through NAFTA, but Mexico refused. We quickly learned that the model doesn't work. I spent many years talking to Mexican officials about the idea of opening up the oil sector to foreign companies. And there have been other efforts to work

toward greater production levels, but it gets mashed in Latin American politics. It's one of these things that's out there, but it quickly runs into ground of economics and geopolitics.

However, American policies toward Latin America, especially Venezuela, are unnecessarily, and arguably mistakenly, heightening foreign investment risks. At this juncture, oil is a common interest, and Chavismo should be irrelevant to the United States.

A Venezuela without Chávez—whether through illness or elections—is not a game changer for the United States. It might be for the anti-Castro community in the United States, but geopolitically his movement peaked already and is no longer a threat. Brazil will remain the model. Chávez is also desperate to reform his Revolution in order to ensure its survival, especially as his health comes into question. With or without Chávez, oil will flow out of Venezuela. How fast will depend a lot on America's relations to Latin America.

Cuba, Chávez, and the ALBA group can't continue to dominate relations. Rather than impose an agenda, Washington needs to really sit down and listen to what countries like Brazil, Argentina, Chile, Peru, Colombia, and Mexico have to say. Latin Americans have other priorities, and what's more, so do most Americans. The economy is and will remain the most pressing problem for all, including Hispanic voters in the United States. Not drugs, not immigration, not Chávez. And every country in the Western Hemisphere shares the goal of increasing oil production as quickly as possible. However you look at it, more energy output and cooperation benefits all, whether it's to contain Chavismo, to energize the American economy, to expand capitalism, or to lift millions more out of poverty.

Obama is missing a window of opportunity. This effort should be supported, not threatened with sanctions by the United States. And whoever occupies the White House should do everything in their power to dispel concerns about Venezuela's oil industry or Latin America's broader stability.

Whatever the United States does about Venezuela, it must stop dragging oil into the diplomatic spat. The world's biggest economy cannot long survive the combination of slow growth and high oil prices, and the standoff with Iran offers enough geopolitical concern as it is. On the other hand, by imposing sanctions or by swaying US public opinion to believe Venezuela is a regime like those in Syria or North Korea, America is only compounding investor fears that Venezuela is a dead end.

The United States needs to cover its blind spot in Latin America, change course, improve ties with the region, and anything else to increase crude output. But it's not. In Trinidad and Tobago in 2009, when he met his counterparts, Obama launched the Energy and Climate Partnership of America. In practice though, the program has achieved little in terms of energy security with its $30 million annual budget. Contrast that to the billions other countries are pouring mostly into oil production, which offer immediate benefits for cash-strapped governments.

Another example is a $2 billion credit line that the US government offered to Petrobras through its Export-Import Bank in 2009. That is only a fraction of what other countries have committed, especially China, but the announcement was still harshly criticized by Republicans who questioned why the US government was lending money to develop offshore oil production in a foreign land and not in the United States. Fair argument, but an erred domestic policy should not translate into more handicaps to raise oil production elsewhere. Besides, Petrobras has only used $300 million, and to that effect, only to underwrite a loan with JP Morgan Chase to purchase American-made oil industry equipment from 150 companies.[1]

These efforts are no match to Venezuelan oil diplomacy, especially with the poorest countries that ultimately end up depending more on Chávez. Even Uribe sees the need for the United States to confront Chávez's oil diplomacy:

> The US and multilateral banks need to make a great effort to help poor countries pay for their oil, to diversify energy resources.

And for that matter, the Arab Spring is certainly a powerful lesson. The United States utterly misread reality on the ground for decades. Middle Eastern dictatorships started wobbling only after their regimes could not economically perpetuate their populist, authoritarian policies. America contributed nothing to those grassroot revolutions, and if anything coexisted with the authoritarian regimes. And yet in Tunisia, Libya, and Egypt, regimes crumbled under their own weight. On the other hand, the Iranian regime that the United States has targeted for years is still resilient.

---

[1]www.exim.gov/brazil/petrobasfacts.cfm

Policymakers have to ask themselves some simple questions: What is the benefit of limiting Venezuela's oil production? Are Cuba, Iran, and North Korea not good examples of how regimes reaffirm their populist hold on power when the United States confronts them? Hasn't Reagan's strategy proved itself a lot more effective than Bush's and Obama's?

Improved relations with Latin America's closest allies will indirectly help Venezuela produce more oil by attracting more investment and diminishing the geopolitical concerns about its oil sector. Doing so will benefit American security in every sector, and in the process consolidate Latin American stability and development based on a pro-market, democratic, and politically centered model. It will also undermine Chavismo regionally, while giving Brazil more time to wear Chávez down. Brazil has more to gain than the United States from a stable Venezuela, and will be more effective than the United States.

Weisbrot, of the CEPR:

> A big change has taken place in Latin America. The whole region has become more independent of the US. South America is more independent than Europe is. The US foreign policy establishment, from government to think tanks and the media, they are looking at this as how the US can restore dominance in the region. They don't look at it as how we'll deal with a new reality. It's like an alcoholic that doesn't realize they have a drinking problem.
>
> They could maintain commercial relations and gain from that, but they are losing economic opportunities because of foreign policy.

The United States can indirectly contain Chávez, and in fact weaken Chávez's hold on power, while increasing oil production and American energy security. In his populist contradictions, Chávez has played the anti-American card, but he has no problem with Brazilian or Chilean models, which are pretty much what the United States wants in Latin America. Furthermore, one of Chávez's closest allies lately is Colombia's President Santos, even if they are ideological antagonists.

Chavismo is already self-contained because it needs foreign investment to survive. With or without Chávez in the picture, with or without socialism, Venezuela will have no choice but to attract foreign investment to its oil sector. How much and how quickly increasingly depends on how much America wants to confront Chávez for petty domestic political gain. The legacy of Chavismo is at stake and its leader will do everything in his power to secure it. Whether it's Chinese, European, or American money, Venezuela will invariably implement more market-friendly policies. To put it simply, confronting Chávez will be as useless as playing nice with him. The best option is to defeat him at his own game.

The United States cannot lag behind Venezuela or China in development aid or soft-credit financing to the region. The United States can't continue to position itself as an indifferent overlord at odds with most of Latin America. The United States can only outmaneuver Chávez if his populist temptation is outbid, and fortunately that can be easily achieved as long as Chávez continues discrediting his own economic model.

A new realist approach could also usher in reforms in policy toward Cuba, at the very least to allow US companies to invest in its promising oil sector. There is room to improve relations with Ecuador, which could be producing more oil with more foreign investment and expertise. Brazilian oil industry would certainly benefit from a new regional policy in the form of more technological cooperation. Brazil is suspicious of the United States and the United States appears indifferent. The result is other foreign players are favored. And if nothing else, the United States has to avoid risk-heightening policies like imposing sanctions on Venezuela, one of the best positioned countries to actually impact world oil prices. It doesn't matter what companies are involved, whether American or Chinese, in pumping out more oil. It doesn't matter where the oil is sent to, whether Japan or Belarus.

Idealist will say the United States cannot budge as democracy is being trampled on by Chávez, that the Communists cannot be left off the hook. But surely the rest of the world cannot all be accomplices to the evils that they denounce. American concerns are shared by its friends, from Brazil and Canada, to Norway and Japan, but they are dealing with Chávez in different ways. And as illustrated by the extreme case of Colombia, and the rest of the continent for that matter, the realist approach has been a lot more effective.

Furthermore, the United States can still pressure Chávez over his human rights record, freedom of press, and so on, the same way it does with other countries like Saudi Arabia, China, Russia, and others. Just leave oil out of it for purely selfish reasons, at the very least.

Of course, even if the United States resorts to simply ignoring Chávez, not sanctioning him and perhaps even seeking more cordial relations, the ability to raise oil output from Venezuela will almost entirely depend on Chávez's domestic oil policies and whether they can attract the capital and expertise to achieve his development goals.

Peñate on the policy alternatives:

> One, the option of confronting Chávez head-on with an embargo (by imposing sanctions) would only strengthen him because the cost would be so high, the population would only side with him. It would send oil prices through the roof and that would only benefit Chávez. He would sell to other markets, perhaps less, but at a higher price. Two, the option of convincing him to change will not work.
>
> Remember, Chávez's success is achieved by American failure in the region. The project is based on more independence [from the US]. Many of us agree with that, although I don't share Chávez's methods. We cannot do that at the expense of private enterprise or democracy. We have to develop an agenda independent of the US because we become more prosperous, more democratic, because the rule of law is strengthened.
>
> The US has no leverage on Venezuela. None. Venezuelan officials don't talk to American officials. Colombia cares about what the US says, but Venezuela doesn't. The Venezuelan government could care less. Even if the US wants to do something about Venezuela, it couldn't, at least playing under civilized rules.
>
> If I advised the US government, I would tell them to resolutely respect the Venezuelan government and the OAS's charter on non-intervention, and not to ever partake in any effort to undermine the Venezuelan regime, or take any military action, covert or overt, legal or illegal. Let him feel he is not a target. I would find the way to let other countries know about this. I would tell Brazil, the EU, and Argentina. I would let Chávez rule out he will be the next Salvador Allende (of Chile).
>
> I would concentrate on containing him. And I would have more concrete soft power programs, instead of FTA promises that aren't met, or embassies that scold local governments because they have problems with counterfeiting, or immigration. I would make all that soft power return, the Peace Corps, campaigns to operate on cataracts, investment campaigns for US businessmen.
>
> It's hard, considering the high fiscal deficit that the US has, but I would try to compete with Venezuela's regional populist programs.

> All it can do is prevent the Chavismo whale from expanding into more countries. But you don't achieve that by promoting a military coup against a Chavista candidate who won the elections. You do that by opening more markets, by introducing an agenda of mutual benefit in the region.
>
> It must be more lucrative for Brazil to do business with the US than with Venezuela, same for El Salvador and other countries, and it often isn't the case.
>
> These are not basket case countries. The US used to look at the region as an investment opportunity. That seems to be disappearing. This is the region where threats come from, not opportunities. The region got the reputation of being the source of bad things, like drugs, immigration, dictators, and wars. But it's no longer a business opportunity.

The United States, like Colombia, could also claim the moral high ground. Chávez would have only himself to blame for not attracting capital and expertise to develop his oil wealth and Bolivarian 21st Century Socialism.

An option would be to rely on Brazil to carry the burden of regional power, much like Saudi Arabia does with Arab nations. If it's a question of results, the United States would do well to support and in fact strengthen the South American approach to dealing with Chávez.

Chávez is far from representing a direct threat to US security. He can sleep easy though. The United States is not going to invade Venezuela anytime soon. But it is unclear—at least in the minds of many Latin American leaders—whether the Obama administration is really moving toward building a case against Chávez. If the Obama administration is targeting Venezuela motivated by electoral gains, its plans will backfire in the form of higher oil prices, a much more damaging political concern for Americans.

Furthermore, using oil as a weapon, especially by increasing its supply, is alien to an administration that has cloaked itself in "green."

Implementing a new approach will invariably take some time, and for that matter, will be delayed until 2013 because the Obama administration is unlikely to review its policies toward the region ahead of elections, nor will Chávez be any more friendly. If anything, expect the standoff to heat up from both sides. His illness will be definitive to bilateral relations in one way or another, although it should be irrelevant to the broader goal of raising oil output from Venezuela.

In any case, bilateral relations with Venezuela and the ALBA group have to be circumscribed to a broader strategic shift. The United States must reassess its paternalistic, idealist regional policy based on domestic issues that the rest of Latin America doesn't share, like the drug war, immigration, and a neoliberal free trade agenda, and negotiate shared priorities. The new approach should be built upon a realist policy that combines pro-market and poverty reduction goals, democracy and balanced free trade, not a value-oriented checklist of carrots and sticks.

In fact, it's in America's interest to potentiate this assertive social-democratic movement to undermine Chavismo, while reclaiming a commanding economic presence in the region. That will, in practice, involve working with Brazil as a regional power, not a junior partner. This would not entail United States sacrificing its interests or even its influence. In fact, it could broaden both.

Brazil and other countries have been demanding it for some time. But this strategic shift should be considered urgent and unavoidable now. Regardless of who is in power in the White House or Congress, both parties have to shift gear quickly in what undoubtedly will be a long process.

Latin America's most important countries and America's allies are already moving in that direction without the United States. The new Latin America is a reality, not a possibility. Whether the United States wants to even notice or deal with it is its choice. But in the meantime, Latin America will continue transitioning toward its new, well-earned position in the world.

Moreover, American regional policy is simply based on an incorrect, or at least untimely, assessment. It's a question of policymakers realizing there is a huge region filled with friends and opportunities, not just trouble. And that this region can be a significant partner in addressing some of America's most pressing issues, especially in the economic front.

The United States has a lot to gain, especially preserving its long-held continental hegemony with more pressing global priorities in mind. The benefits will translate into American jobs, American exports, and American economic growth; it will further consolidate the political evolution of Latin America toward a democratic, developed region, especially in smaller, weaker countries unable to combat poverty on their own; it will reaffirm geopolitical, strategic, and cultural ties, as well as facilitate addressing pending domestic issues like immigration and the drug war. And most importantly for the United States, it is sure to improve American energy and national security one way or another in this troubling decade ahead.

There is little room for debate. Improving relations with Latin America will yield a significant net positive return for both sides in just about everything. The debate is solely regarding how to go about it. And considering that the idealist policies of the past two decades have failed to expand or even protect America's position in the region, a new approach is long overdue.

It's time to cover America's blind spot.

INDEX